Mobile Opportunistic Networks

Architectures, Protocols and Applications

Mobile Opportunistic Networks

Architectures, Protocols and Applications

Edited by Mieso K. Denko

CRC Press
Taylor & Francis Group
Boca Raton London New York

CRC Press is an imprint of the
Taylor & Francis Group, an **informa** business

AN AUERBACH BOOK

CRC Press
Taylor & Francis Group
6000 Broken Sound Parkway NW, Suite 300
Boca Raton, FL 33487-2742

First issued in paperback 2019

ISBN-13: 978-1-4200-8812-0 (hbk)
ISBN-13: 978-0-367-38268-1 (pbk)

Library of Congress Cataloging-in-Publication Data

Mobile opportunistic networks : architectures, protocols and applications / editor, Mieso K. Denko.
        p. cm.
    Includes bibliographical references and index.
    ISBN 978-1-4200-8812-0 (hardcover : alk. paper)
    1. Wireless communication systems. 2. Mobile computing. 3. Computer networks. I. Denko, Mieso K.

TK5103.2.M592 2011
004.6--dc22   2010022641

Visit the Taylor & Francis Web site at
http://www.taylorandfrancis.com

and the CRC Press Web site at
http://www.crcpress.com

Contents

Preface

Opportunistic networks are an emerging networking paradigm where communication between the source and destination happens on the fly and depends on the availability of communication links. In opportunistic networks, intermittent connectivity is frequent and mobile nodes can communicate with each other even if a route connecting them did not previously exist. In this type of network, it is not mandatory to have a priori knowledge about the network topology. The network is formed opportunistically based on proximity and network availability, by randomly connecting and disconnecting the networks and devices. This networking paradigm heavily benefits from the heterogeneous networking and communication technologies that currently exist and will emerge in the future. Hence, given the advances in wireless networking technologies and the wide availability of pervasive and mobile devices, opportunistic network applications are promising networking and communication technologies for a variety of future mobile applications. Mobile opportunistic networks introduce several research challenges in all aspects of computing, networking, and communication.

This book provides state-of-the-art research and future trends in mobile opportunistic networking and applications. The chapters, contributed by prominent researchers from academia and industry, will serve as a technical guide and reference material for engineers, scientists, practitioners, graduate students, and researchers. To the best of my knowledge, this is the first book on mobile opportunistic networking.

The book is organized into two sections covering diverse topics by presenting state-of-the-art architectures, protocols, and applications in opportunistic networks.

Section 1: Architectures and Protocols

Section 1 consists of Chapters 1 through 6, which focus on modeling, networking architecture, and routing problems in opportunistic networking.

Chapter 1, *Routing in Mobile Opportunistic Networks*, is by Libo Song and David F. Kotz, and discusses routing in mobile opportunistic networks. The simulation of several routing protocols in opportunistic networks are evaluated and discussed. The authors have also presented and evaluated their proposed prediction-based routing protocol for opportunistic networks. This protocol was evaluated using realistic contact traces, and then compared with existing routing protocols. Chapter 2, *State of the Art in Modeling Opportunistic Networks,* was written by Thabotharan Kathiravelu and Arnold Pears, and discusses the state of the art in modeling opportunistic network connection structures and pairwise contacts. The chapter also introduces connectivity models as an approach to modeling contacts in opportunistic networks, and then illustrates the scope of this approach using case studies. Chapter 3 is entitled *Credit-Based Cooperation Enforcement Schemes Tailored to Opportunistic Networks,* and was written by Isaac Woungang and Mieso K. Denko. This chapter discusses cooperation enforcement in opportunistic networks. A comprehensive review and detailed comparison of credit-based incentive schemes for mobile wireless ad hoc networks and challenged networks are presented, with the goal of identifying those that are tailored to opportunistic networks. Chapter 4, *Opportunism in Mobile Ad Hoc Networking* by Marcello Caleffi and Luigi Paura, discusses some fundamental characteristics of opportunistic routing. Most of the popular existing routing protocols and their unique features and suitability to mobile opportunistic networks are discussed. Chapter 5, *Opportunistic Routing for Load Balancing and Reliable Data Dissemination in Wireless Sensor Networks* by Min Chen, Wen Ji, Xiaofei Wang, Wei Cai, and Lingxia Liao, proposes a novel opportunistic routing protocol for delivering data with load balancing and reliable transmission capabilities. Performance results in terms of network lifetime and transmission reliability are discussed. Chapter 6 is entitled *Trace-Based Analysis of Mobile User Behaviors for Opportunistic Networks,* and is written by Wei-Jen Hsu and Ahmed Helmy. This chapter presents a framework that provides a procedural approach to analyzing user behavior based on realistic data in opportunistic networks. The authors have employed a data-driven approach to develop a fundamental understanding of realistic user behavior in mobile opportunistic networks.

Section 2: Services and Applications

Section 2 consists of Chapter 7 through 10 and focuses on opportunistic networking technologies and applications.

Chapter 7, *Quality of Service in an Opportunistic Capability Utilization Network* by Leszek Lilien, Zille Huma Kamal, Ajay Gupta, Isaac Woungang, and Elvira Bonilla Tamez, presents opportunistic networks (Oppnets) as a paradigm and a technology proposed for realization of opportunistic capability utilization networks. This chapter also presents the use of the service location and planning (SLP) technique for resource utilization with quality of service (QoS) constraints in opportunistic capability utilization networks. It also illustrates the use of Semantic Web technology and its ontologies for specifying QoS requirements in Oppnets using a novel Oppnet model. Chapter 8, *Effective File Transfer in Mobile Opportunistic Networks* by Ling-Jyh Chen and Ting-Kai Huang, presents a peer-to-peer approach for mobile file transfer applications in opportunistic networks. This chapter also discusses the combined strengths of cache-based approaches and Infostation-based approaches, as well as the implementation of a collaborative forwarding algorithm to further utilize opportunistic ad hoc connections and spare storage in the network. Chapter 9, *Stationary Relay Nodes Deployment on Vehicular Opportunistic Networks* by Joel J. P. C. Rodrigues, Vasco N. G. J. Soares, and Farid Farahmand, reviews recent advances in the deployment of stationary relay nodes on vehicular opportunistic networks. This chapter also discusses the impact of adding stationary relay nodes on the performance of delay-tolerant network routing protocols as applied to vehicular opportunistic networks. Finally, Chapter 10, *Connection Enhancement for Mobile Opportunistic Networks* by Weihuang Fu, Kuheli Louha, and Dharma P. Agrawal, presents analytical approaches for mobility and heterogeneous connections management in mobile opportunistic networks. Strategies are introduced for network connection selection and message forwarding based on the author's analytical work. The authors also analyze the improvement of heterogeneous connections for message delivery performance and have presented a detailed investigation of the current state-of-the-art protocols and algorithms.

The research in mobile opportunistic computing and networking is currently in progress in academia and industry. Although this book may not be an exhaustive representation of all research efforts in the area, they do represent a good sample of key aspects and research trends.

We owe our deepest gratitude to all the authors for their valuable contributions to this book and their great efforts and cooperation. We wish to express our thanks to Auerbach Publications, the CRC Press staff, and especially to Rich O'Hanley and Stephanie Morkert for their excellent guidance and support.

Finally, I would like to dedicate this book to my wife Hana and our children for their support and understanding throughout this project.

Dr. Mieso K. Denko
November 2009

About the Editor

Mieso K. Denko received his MSc degree from the University of Wales, United Kingdom and his PhD degree from the University of Natal, South Africa, both in Computer Science. He is a founding Director of the Pervasive and Wireless Networking Research Lab in the Department of Computing and Information Science, University of Guelph, Ontario, Canada. His current research interests include wireless networks, mobile and pervasive computing, wireless mesh networks, wireless sensor networks, and network security. His research results in these areas have been published, in international journals, in conference proceedings, and contributed to books. Dr. Denko is a founder/co-founder of a number of ongoing international workshops and symposia. He has served on several international conferences and workshops as general vice-chair, program co-chair/vice-chair, publicity chair, and technical program committee member. He has guest co-edited several journal special issues in Springer, Wiley, Elsevier, and other journals. Most recently he guest co-edited journal special issues in ACM/Springer Mobile Networks and Applications (MONET) and IEEE Systems Journal (ISJ). Dr. Denko has edited/co-edited multiple books in the areas of pervasive and mobile computing, wireless networks, and autonomic networks. Most recently he co-edited two books: *Wireless Quality of Service: Techniques, Standards, and Applications,* published by Auerbach Publications, September 2008, and *Autonomic Computing and Networking,* published by Springer in June 2009. He is Associate Editor of international journals, including the *International Journal of Communication Systems* (Wiley), the *Journal of Ambient Intelligence and Humanized Computing* (Springer), and *Security and Communication Networks Journal* (Wiley). Dr. Denko is a senior member of the ACM and IEEE and the Vice Chair of IFIP WG 6.9.

He passed away in April 2010.

Chapter 1

Routing in Mobile Opportunistic Networks

Libo Song
Google, Inc

David F. Kotz
Dartmouth College

Contents

1.1 Routing in Mobile Opportunistic Networks*

Routing in mobile ad hoc networks has been studied extensively. Most of these studies assume that a contemporaneous end-to-end path exists for the network nodes. Some mobile ad hoc networks, however, may not satisfy this assumption. In mobile sensor networks [26], sensor nodes may turn off to preserve power. In wild-animal tracking networks [13], animals may roam far away from each other. Other networks, such as pocket-switched networks [9], battlefield networks [7,18], and transportation networks [1,16], may experience similar disconnections due to mobility, node failure, or power-saving efforts.

One solution for message delivery in such networks is that the source passively waits for the destination to be in communication range and then delivers the message. Another active solution is to flood the message to any nodes in communication range. The receiving nodes carry the message and repeatedly flood the network with the message. Both solutions have obvious advantages and disadvantages: the first may have a low delivery ratio while using few resources; the second may have a high delivery ratio while using many resources.

Many other opportunistic routing protocols have been proposed in the literature. Few of them, however, were evaluated in realistic network settings, or even in realistic simulations, due to the lack of any realistic people mobility model. Random walk or random way-point mobility models are often used to evaluate the performance of those routing protocols. Although these synthetic mobility models have received extensive interest from mobile ad hoc network researchers [2], they do not reflect people's mobility patterns [10]. Realizing the limitations of using random mobility models in simulations, a few researchers have studied routing protocols in mobile opportunistic networks with realistic mobility traces. In the Haggle project, Chaintreau et al. [3] theoretically analyzed the impact of routing algorithms over a model derived from a realistic mobility data set. Su et al. [23] simulated a set of routing protocols in a small experimental network. Those studies help researchers better understand the theoretical limits of opportunistic networks and the routing protocol performance in a small network (20–30 nodes).

Deploying and experimenting with large-scale mobile opportunistic networks is difficult, so we also resort to simulation. Instead of using a complex mobility

---

* This work is based on an earlier work: "Evaluating Opportunistic Routing Protocols with Large Realistic Contact Traces," in Proceedings of the Second ACM Workshop on Challenged Networks, ©ACM 2007. http://doi.acm.org/10.1145/1287791.1287799.

model to mimic people's mobility patterns, we used mobility traces collected in a production wireless network at Dartmouth College to drive our simulation. Our message-generation model, however, was synthetic.

We study protocols for routing messages between wireless networking devices carried by people. We assume that people send messages to other people occasionally, using their devices; when no direct link exists between the source and the destination of the message, other nodes may relay the message to the destination. Each device represents a unique person (it is out of the scope of our work to cover instances when a device may be carried by different people at different times). Each message is destined for a specific person and thus for a specific node carried by that person. Although one person may carry multiple devices, we assume that the sender knows which device is the best to receive the message. We do not consider multicast or geocast in this chapter.

Using realistic contact traces, which we derived from the Dartmouth College dataset [15], we evaluated the performance of three "naive" routing protocols (direct-delivery, epidemic, and random) and two prediction-based routing protocols, PRoPHET [19] and Link-State [23]. We also propose a new prediction-based routing protocol and compare it to the above protocols.

1.1.1 Routing Protocol

A routing protocol is designed for forwarding messages from one node (source) to another node (destination). Any node may generate messages for any other node and may carry messages destined for other nodes. We consider only messages that are unicast (single destination).

Delay-tolerant networks (DTN) routing protocols can be described in part by their *transfer probability* and *replication probability*; that is, when one node meets another node, what is the probability that a message should be transferred and, if so, whether the sender should retain its copy. Two extremes are the direct-delivery protocol and the epidemic protocol. The former transfers with probability 1 when the node meets the destination, 0 for others, and never replicates a packet; in effect, the packet only moves when the source meets the destination. The latter uses transfer probability 1 for all nodes and replicates the packet each time it meets another node. Both these protocols have their advantages and disadvantages. All other protocols are between the two extremes.

First, we define the notion of contact between two nodes. Then we describe five existing protocols before presenting our own proposal.

A *contact* is defined as a period of time during which two nodes have the opportunity to communicate. Although wireless technologies differ, we assume that a node can reliably detect the beginning and end time of a contact with nearby nodes. A node may be in contact with several other nodes at the same time.

The contact history of a node is a sequence of contacts with other nodes. Node i has a contact history H_{ij} for each other node j, which denotes the historical contacts between node i and node j. We record the start and end time for each contact; however, the last contacts in the node's contact history may not have ended.

1.1.1.1 Direct Delivery Protocol

In this simple protocol, a message is transmitted only when the source node can directly communicate with the destination node of the message. In mobile opportunistic networks, however, the probability for the sender to meet the destination may be low, or even zero.

1.1.1.2 Epidemic Routing Protocol

The epidemic routing protocol [24] floods messages into the network. The source node sends a copy of the message to every node that it meets. The nodes that receive a copy of the message also send a copy of the message to every node that they meet. Eventually, a copy of the message arrives at the destination of the message.

This protocol is simple but may use significant resources; excessive communication may drain each node's battery quickly. Moreover, since each node keeps a copy of each message, storage is not used efficiently and the capacity of the network is limited.

At a minimum, each node must expire messages after some amount of time or stop forwarding them after a certain number of hops. After a message expires, the message will not be transmitted and will be deleted from the storage of any node that holds the message.

An optimization to reduce the communication cost is to transfer *index messages* before transferring any data message. The index messages contain IDs of messages that a node currently holds. Thus, by examining the index messages, a node only transfers messages that are not yet contained on the other nodes.

1.1.1.3 Random Routing

An obvious approach between the two extremes previously discussed is to select a transfer probability between 0 and 1 to forward messages at each contact. The replication factor can also be a probability between 0 (none) and 1 (all). For our random protocol, we use a simple replication strategy that makes no replicas. The message is transferred every time the transfer probability is greater than the threshold. The message has some chance of being transferred to a highly mobile node and thus may have a better chance to reach its destination before the message expires.

1.1.1.4 PRoPHET Protocol

PRoPHET [19] is the Probabilistic Routing Protocol using History of past Encounters and Transitivity, which is used to estimate each node's delivery probability for each other node. When node i meets node j, the delivery probability of node i for j is updated by

$$p'_{ij} = (1 - p_{ij})p_0 + p_{ij}, \tag{1.1}$$

where p_0 is an initial probability, a design parameter for a given network. Lindgren et al. [19] chose 0.75, as did we in our evaluation. When node i does not meet j for some time, the delivery probability decreases by

$$p'_{ij} = \alpha^k p_{ij},$$

(1.2)

where α is the aging factor ($\alpha < 1$), and k is the number of time units since the last update.

The PRoPHET protocol exchanges index messages as well as delivery probabilities. When node i receives node j's delivery probabilities, node i may compute the transitive delivery probability through j to z with

$$p'_{iz} = p_{iz} + (1 - p_{iz}) p_{ij} p_{jz} \beta,$$

(1.3)

where β is a design parameter for the impact of transitivity; we use $\beta = 0.25$ (as in [19]).

ZebraNet [13] used a simple history-based routing protocol, which calculates the probability similar to Equation (1.2) to estimate the probability that a node can communicate with the destination, the base station. Only one base station exists in ZebraNet.

1.1.1.5 Link-State Protocol

Su et al. [23] use a link-state approach to estimate the weight of each path from the source of a message to the destination. They use the median intercontact duration or exponentially aged intercontact duration as the weight on links. The exponentially aged intercontact duration of node i and j is computed by

$$w'_{ij} = \alpha w_{ij} + (1 - \alpha)(t - t_{ij}),$$

(1.4)

where t is the current time, t_{ij} is the time of last contact, and α is the aging factor. At the first contact, we just record the contact time.

Nodes share their link-state weights when they can communicate with each other, and messages are forwarded to the neighbor that has the path to destination with the lowest link-state weight.

1.1.2 Timely Contact Probability

We, too, use historical contact information to estimate the probability of meeting other nodes in the future. But our method differs in that we estimate the contact probability within a period of time. For example, what is the contact probability in the next hour? Neither PRoPHET nor Link-State considers time in this way.

One way to estimate the *timely contact probability* is to use the ratio of the total contact duration to the total time. However, this approach does not capture the frequency of contacts. For example, one node may have a long contact with another node, followed by a long noncontact period. A third node may have a short contact with the first node, followed by a short noncontact period. Using the above estimation approach, both examples would have similar contact probability. In the second example, however, the two nodes have more frequent contacts.

We design a method to capture the contact frequency of mobile nodes. For this purpose, we assume that even short contacts are sufficient to exchange messages.*

The probability for node i to meet node j is computed by the following procedure. We divide the contact history H_{ij} into a sequence of n periods of ΔT starting from the start time (t_0) of the first contact in history H_{ij} to the current time. We number each of the n periods from 0 to $n-1$, then check each period. If node i had any contact with node j during a given period m, which is $[t_0 + m\Delta T, t_0 + (m+1)\Delta T)$, we set the contact status I_m to be 1; otherwise, the contact status I_m is 0. The probability $p_{ij}^{(0)}$ that node i meets node j in the next ΔT can be estimated as the average of the contact status in prior intervals:

$$p_{ij}^{(0)} = \frac{1}{n} \sum_{m=0}^{n-1} I_m. \qquad (1.5)$$

To adapt to the change of contact patterns, and reduce the storage space for contact histories, a node may discard old contacts from the history; in this situation, the estimate would be based on only the retained history.

The above probability is the direct contact probability of two nodes. We are also interested in the probability that we may be able to pass a message through a sequence of k nodes. Therefore, we need not only use the node with highest contact probability, but also several other nodes. With all nodes' contact probabilities, we can compute the k-order probability. Assuming nodes' contact events are independent, we define the k-order probability inductively,

$$p_{ij}^{(k)} = p_{ij}^{(0)} + \sum_{r} p_{ir}^{(0)} p_{rj}^{(k-1)}, \qquad (1.6)$$

where r is any node other than i or j. Since node contact events may not be entirely independent, we recognize that Equation (1.5) only estimates an *approximate* probability. Indeed, all of the prediction-based methods use heuristic approximations and incomplete data, given the nature of this routing problem.

1.1.3 Our Routing Protocol

We first consider the case of a two-hop path—that is, with only one relay node. We consider two approaches: either the receiving neighbor decides whether to act as a relay, or the source decides which neighbors to use as relays.

---

* In our simulation, however, we accurately model the communication costs and some short contacts will not succeed in the transfer of all messages.

1.1.3.1 Receiver Decision

Whenever a node meets other nodes, they exchange all their messages (or, as above, index messages). If the destination of a message is the receiver itself, the message is delivered. Otherwise, if the probability of delivering the message to its destination through this receiver node within ΔT is greater than or equal to a certain threshold, the message is stored in the receiver's storage to forward to the destination. If the probability is less than the threshold, the receiver discards the message. Notice that our protocol replicates the message whenever a good relay comes along.

1.1.3.2 Sender Decision

To make decisions, a sender must have the information about its neighbors' contact probability with a message's destination. Therefore, meta-data exchange is necessary.

When two nodes meet, they exchange a meta-message, containing an unordered list of node IDs for which the sender of the meta-message has a contact probability greater than the threshold.

After receiving a meta-message, a node checks whether it has any message that is destined to its neighbor or to a node in the node list of the neighbor's meta-message. If it has, it sends a copy of the message.

When a node receives a message, if the destination of the message is the receiver itself, the message is delivered. Otherwise, the message is stored in the receiver's storage for forwarding to the destination.

1.1.3.3 Multinode Relay

When we use more than two hops to relay a message, each node needs to know the contact probabilities along all possible paths to the message destination.

Every node keeps a contact probability matrix in which each cell p_{ij} is a contact probability between two nodes i and j. Each node i computes its own contact probabilities (row i) using Equation (1.5) whenever the node ends a contact with other nodes. Each row of the contact probability matrix has a version number; the version number for row i is increased only when node i updates the matrix entries in row i. Other matrix entries are updated through exchange with other nodes when they meet.

When two nodes i and j meet, they first exchange their contact probability matrices. Node i compares its own contact matrix with node j's matrix. If node j's matrix has a row l with a higher version number, then node i replaces its own row l with node j's row l. Likewise, node j updates its matrix. After the exchange, the two nodes will have identical contact probability matrices.

Next, if a node has a message to forward, the node estimates its neighboring node's order-k contact probability to contact the destination of the message using Equation (1.6). The order k is a design factor for the multinode relay protocols. If

$p_{ij}^{(m)}$ for any $0 < m < k$, is above a threshold, or if j is the destination of the message, node i will send a copy of the message to node j.

All the previous effort serves to determine the transfer probability when two nodes meet. The replication decision is orthogonal to the transfer decision. In our implementation, we always replicate. Although PRoPHET [19] and Link-State [23] do no replication, as described, we added replication to both protocols for better comparison to our protocol.

1.1.4 Evaluation Results

We evaluate and compare the results of direct delivery, epidemic, random, PRoPHET, Link-State, and timely contact routing protocols.

1.1.4.1 Mobility Traces

We use real mobility data collected at Dartmouth College. Dartmouth College has collected association and disassociation messages from devices on its wireless network since spring 2001 [15]. Each message records the wireless card MAC address, the time of association and disassociation, and the name of the access point (AP). We treat each unique MAC address as a node. For more information about Dartmouth's network and the data collection, see previous studies [8,14].

Our data are not contacts in a mobile ad hoc network. We can approximate contact traces by assuming that two users can communicate with each other whenever they are associated with the same access point. Chaintreau et al. [3] used Dartmouth data traces and made the same assumption to theoretically analyze the impact of human mobility on opportunistic forwarding algorithms. This assumption may not be accurate,* but it is a good first approximation. In our simulation, we imagine the same clients and same mobility in a network with no access points. Since our campus has full Wi-Fi coverage, we assume that the location of access points had little impact on users' mobility.

We simulated one full month of trace data (November 2003), with 5,142 users. Although prediction-based protocols require prior contact history to estimate each node's delivery probability, our preliminary results show that the performance improvement of warming-up over one month of trace was marginal. Therefore, for simplicity, we show the results of all protocols without warm-up.

---

* Two nodes may not have been able to directly communicate while they were at far sides of an access point, or two nodes may have been able to directly communicate if they were between two adjacent access points.

1.1.4.2 Simulator

We developed a custom simulator.* Since we used contact traces derived from real mobility data, we did not need a mobility model and omitted physical and link-layer details for node discovery. We are aware that the time for neighbor discovery in different wireless technologies varies from less than one second to several seconds. Furthermore, connection establishment also takes time, such as Dynamic Host Configuration Protocol (DHCP). In our simulation, we assumed that the nodes could discover and connect to each other instantly when they were associated with the same AP. To accurately model communication costs, however, we simulated some MAC-layer behaviors, such as collision.

The default settings of the network of our simulator are listed in Table 1.1, using the values recommended by other papers [23,19]. The message probability was the probability of generating messages, as described below in Section 1.1.4.3. The default transmission bandwidth was 11 Mb/s. When one node tried to transmit a message, it first checked whether another node was transmitting. If it was, the node backed off for a random number of slots. Each slot was 1 millisecond, and the maximum number of back-off slots was 30. The size of messages was uniformly distributed between 80 bytes and 1024 bytes. The hop count limit (HCL) was the maximum number of hops before a message should stop forwarding. The time to live (TTL) was the maximum duration that a message may exist before expiring. The storage capacity was the maximum space that a node can use for storing messages. For our routing method, we used a default prediction window ΔT of 10 hours and a probability threshold of 0.01. The replication factor r was not limited by default, so the source of a message transferred the messages to any other node that had a contact probability with the message destination higher than the probability threshold. All protocols (except direct and random) used the *index*-message optimization method above.

1.1.4.3 Message Generation

After each contact event in the contact trace, we generated a message with a given probability; we chose a source node and a destination node randomly using a uniform distribution across nodes seen in the contact trace up to the current time. When there were more contacts during a certain period, there was a higher likelihood that a new message was generated in that period. This correlation is not

---

* We tried to use a general network simulator (ns2), which was extremely slow when simulating a large number of mobile nodes (in our case, more than 5000 nodes) and provided unnecessary detail in modeling lower-level network protocols.

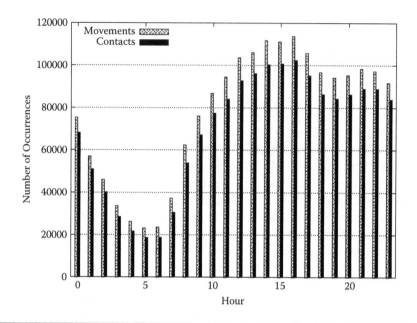

Figure 1.1   Movements and contacts during each hour.

unreasonable, since there were more movements and contacts during the day than during the night. Figure 1.1 shows the statistics of the numbers of movements and the numbers of contacts during each hour. These statistics were accumulated across all users through the whole month. The plot shows a clear diurnal activity pattern; the activities were lowest around 5 a.m. and peaked between 4 p.m. and 5 p.m. We assume that in some applications, network traffic exhibits similar patterns; that is, people send more messages during the day, too.

1.1.4.4 Metrics

We define a set of metrics that we used in evaluating routing protocols in opportunistic networks:

- *Delivery ratio*, the ratio of the number of messages delivered to the number of total messages generated.
- *Delay*, the duration between a message's generation time and the message's delivery time.
- *Message transmissions*, the total number of messages transmitted during the simulation across all nodes.
- *Meta-data transmissions*, the total number of meta-data units transmitted during the simulation across all nodes.

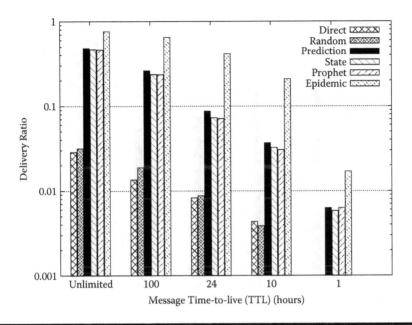

Figure 1.2 Delivery ratio (log scale). The direct and random protocols for one-hour TTL had delivery ratios that were too low to be shown in the plot.

- *Message duplications*, the number of times a message copy occurred.
- *Storage usage*, the max and mean of maximum storage (bytes) used across all nodes.

1.1.4.5 Results

Here we compare our simulation results of the six routing protocols.

Figure 1.2 shows the delivery ratio of all the protocols, with different TTLs. (In all the plots in this section, *prediction* stands for our method, *state* stands for the Link-State protocol, and *prophet* represents PRoPHET.) Although we had 5142 users in the network, the direct-delivery and random protocols had low delivery ratios (note the log scale). Even for messages with an unlimited lifetime, only 59 out of 2077 messages were delivered during this one-month simulation. The delivery ratio of epidemic routing was the best. The three prediction-based routing schemes had low delivery ratios, compared with epidemic routing. Although our method was slightly better than the other two, the advantage was marginal. Note that with a 10-hour TTL, the three prediction-based routing protocols had only about 4% messages delivered. This low delivery ratio limits the applicability of these routing protocols in practice.

The high delivery ratio of epidemic routing came with a price: excessive transmissions. Figure 1.3 shows the number of message data transmissions. The number

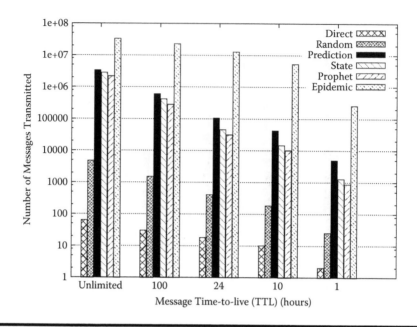

Figure 1.3   Message transmissions (log scale).

of message transmissions in epidemic routing was more than 10 times higher than for the prediction-based routing protocols. Obviously, the direct delivery protocol had the lowest number of message transmissions—the number of messages delivered. Among the three prediction-based methods, PRoPHET transmitted fewer messages, but all three had comparable delivery ratios, as seen in Figure 1.2.

Figure 1.4 shows that epidemic and all prediction-based methods had substantial meta-data transmissions, though epidemic routing had relatively more (at least for shorter TTLs). Because the epidemic protocol transmitted messages at every contact, in turn, more nodes had messages that required meta-data transmission during contact. The direct-delivery and random protocols had no meta-data transmissions.

In addition to its message transmissions and meta-data transmissions, the epidemic routing protocol also had excessive message duplications, spreading replicas of messages over the network. Figure 1.5 shows that epidemic routing had one or two orders of magnitude more duplication than the prediction-based protocols. Recall that the direct-delivery and random protocols did not replicate, and thus had no data duplications.

Figure 1.6 shows the median delivery delays, and Figure 1.7 shows the mean delivery delays. All protocols show similar delivery delays in both mean and median measures for medium TTLs but differ for long and short TTLs. With a 100-hour TTL, or unlimited TTL, epidemic routing had the shortest delays. Direct delivery had the longest delay for unlimited TTL, but it had the shortest delay for the one-hour TTL.

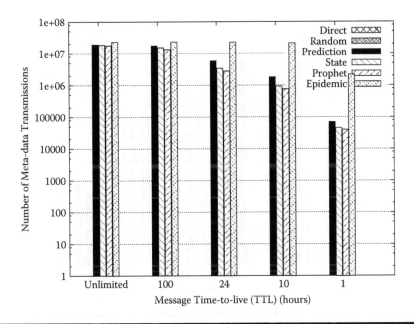

Figure 1.4 Meta-data transmissions (log scale). Direct and random protocols had no meta-data transmissions.

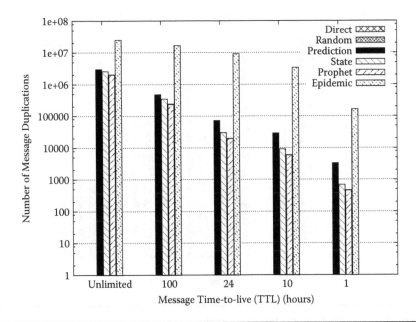

Figure 1.5 Message duplications (log scale). Direct and random protocols had no message duplications.

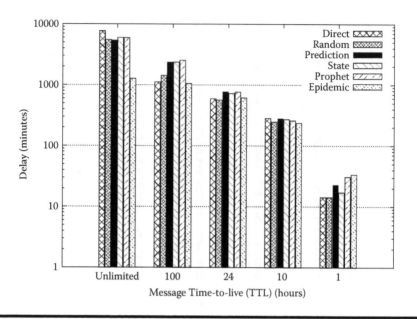

Figure 1.6   Median delivery delay (log scale).

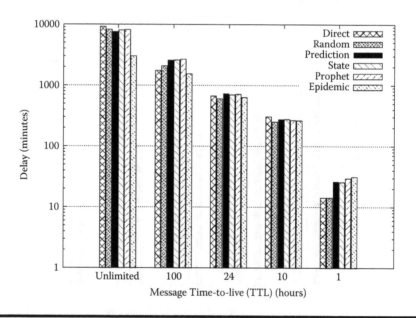

Figure 1.7   Mean delivery delay (log scale).

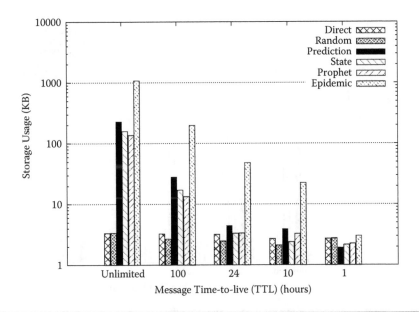

Figure 1.8   Max of maximum storage usage across all nodes (log scale).

The results seem contrary to our intuition: the epidemic routing protocol should be the fastest routing protocol since it spreads messages all over the network. Indeed, the figures show only the delay time for *delivered* messages. For direct delivery, random, and the probability-based routing protocols, relatively few messages were delivered for short TTLs, so many messages expired before they could reach their destination. Those messages had infinite delivery delay and were not included in the median or mean measurements. For longer TTLs, more messages were delivered even for the direct-delivery protocol. The statistics of longer TTLs for comparison are more meaningful than those of short TTLs.

Since our message generation rate was low, the storage usage was also low in our simulation. Figure 1.8 and Figure 1.9 show the maximum and average of maximum volume (in kb) of messages stored in each node. The epidemic routing had the most storage usage. The message time-to-live parameter was the big factor affecting the storage usage for epidemic and prediction-based routing protocols.

We studied the impact of different parameters of our prediction-based routing protocol. Our prediction-based protocol was sensitive to several parameters, such as the probability threshold. Figure 1.10 shows the delivery ratios when we used different probability thresholds. (The leftmost value, 0.01, is the value used for the other plots.) A higher probability threshold limited the transfer probability, so fewer messages were delivered. We also had fewer transmissions, as shown in Figure 1.11. With a larger prediction window ΔT, we got higher contact probability. Thus, for the same probability threshold, we had a higher delivery ratio (Figure 1.12), and

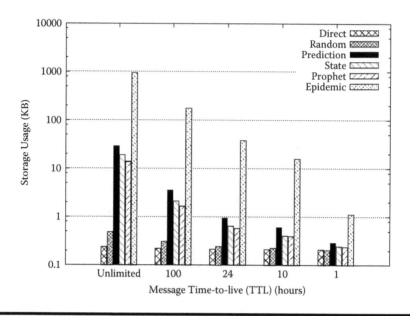

Figure 1.9 Mean of maximum storage usage across all nodes (log scale).

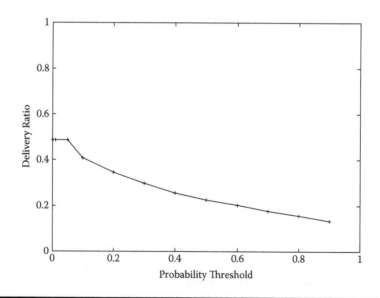

Figure 1.10 Probability threshold impact on delivery ratio of timely contact routing. We used threshold 0.01 for other plots.

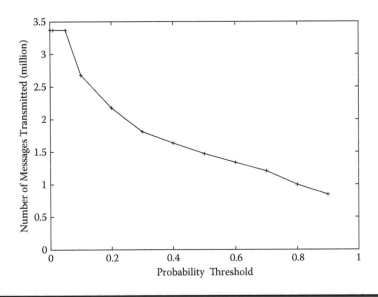

Figure 1.11 **Probability threshold impact on message transmission of timely contact routing. We used threshold 0.01 for other plots.**

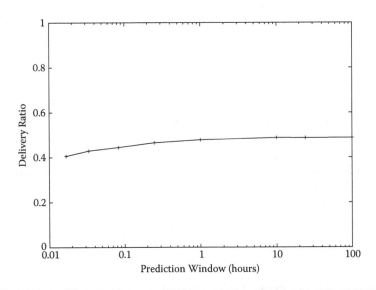

Figure 1.12 **Prediction window impact on delivery ratio of timely contact routing (semi-log scale). We used prediction window $\Delta T = 10$ hours for other plots.**

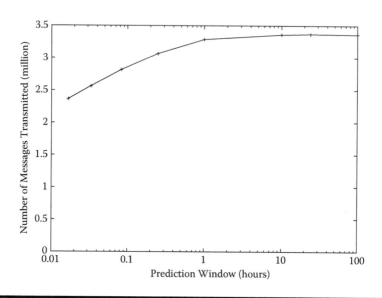

Figure 1.13 **Prediction window impact on message transmission of timely contact routing. We used prediction window $\Delta T = 10$ hours for other plots.**

more transmissions (Figure 1.13). Our protocol was not as sensitive to the prediction window as it was to probability threshold. When the prediction window increased to one hour or longer, the delivery ratio and the number of messages transmitted increased only slightly. Therefore, the contact probability within one hour or longer did not change much.

1.1.5 Related Work

Fall [6] presents an overview of DTNs. It gives examples of delay-tolerant networks: terrestrial mobile networks, exotic media networks (e.g., satellite, deep space), military ad-hoc networks, and sensor networks. The challenges of those networks are high latency, disconnection, and long queuing time. Fall focuses on the interoperability of heterogeneous networks (e.g., Internet, satellite networks, Intranet, or ad-hoc networks). The author proposes to use a DTN gateway to connect different networks, and defines naming of network entities.

Chuah et al. [4] extends the naming convention in Fall's DTN architecture framework [6] to further divide a network region into groups. This work also discusses details of neighbor discovery, gateway selection, mobility management, and route discovery.

Jain et al. [12] later extend Fall's framework [6]. They propose and evaluate several routing algorithms in two example scenarios: remote village and city bus. Both scenarios have predictable connectivity. The remote village has a dial-up connection

Table 1.1   Default Settings of the Opportunistic Network Simulation

Parameter	Default Value
message probability	0.001
bandwidth	11 Mb/s
transmission slot	1 millisecond
max back-off slots	30
message size	80–1024 bytes
hop count limit (HCL)	unlimited
time to live (TTL)	unlimited
storage capacity	unlimited
predictions window ΔT	10 hours
probability threshold	0.01
contact history length	20
replication	always
aging factor α	0.9 (0.98 PRoPHET)
initial probability p_0	0.75 (PRoPHET)
transitivity impact β	0.25 (PRoPHET)

late at night, satellite connection every few hours, and a motorbike message courier every 4 hours. The city buses follow bus routes. Five routing algorithms are evaluated in the paper:

- First Contact (FC): a message is forwarded to one random current contact or to the first available contact.
- Minimum Expected Delay (MED): use Dijkstra's algorithm to compute the cost (delay) of each path and choose an edge that leads to the path that has the least sum of the average waiting time, propagation delay, and transmission delay.
- Earliest Delivery (ED): use modified Dijkstra with time varying cost, but without queue waiting time.
- Earliest Delivery with Local Queue (EDLQ): ED with the cost function incorporating local queuing.
- Earliest Delivery with All Queue (EDAQ): ED with the cost function incorporating all nodes' queuing info.
- Linear Program (LP): use all the contacts, queuing, and traffic information.

They conclude that in networks with limited resources, the smarter algorithms (ED, EDLQ, and EDAQ) may provide a significant benefit. They also found that global knowledge may not be required for good performance, since EDLQ performed as well as EDAQ in many cases. The routing problem, however, when the contacts are predictable, is relatively simple, especially in the remote village scenario, where the village communicates with only the city through three different routes. The LP algorithm provides only a theoretical analysis. In practice, the required information is not available.

Li and Rus [18] propose algorithms to guide mobile nodes' movements for communication in disconnected mobile ad hoc networks. Two algorithms were studied. The first assumes that mobilities and locations are known to all nodes. The second one is more generalized and does not assume this knowledge. Li and Rus describe a network situation in which nodes are moving with their tasks and may move to other nodes for message delivery. The proposed algorithms avoid traditional waiting and retry schemes. The big disadvantage is that the algorithms require users to move based on the needs of messages. This means that users (people) need to be aware of the messages and decide if relay is necessary. The algorithms are tools to aid the user to make the movement decision.

Wang and Wu [26] studied two basic routing approaches (direct delivery and flooding) and propose a scheme based on delivery probability. They found that their delivery-probability scheme achieves a higher delivery ratio than either of the basic routing approaches. They also studied the average delay and transmission overhead for all three delivery methods. Although the flooding approach has the lowest average delay, the delivery ratio suffers because excessive message duplications exhaust storage buffers. The authors discuss a mobile sensor network in which all sensor nodes transmit messages to sink nodes. Messages can be delivered to any sink node. The limited number of destinations enables the simple probability estimation of the proposed scheme.

LeBrun et al. [16] propose a location-based, delay-tolerant network routing protocol. Their algorithm assumes that every node knows its own position, and the single destination is stationary at a known location. A node forwards data to a neighbor only if the neighbor is closer to the destination than its own position. Our protocol does not require knowledge of the nodes' locations, and it learns their contact patterns.

Leguay et al. [17] use a high-dimensional space to represent a mobility pattern, then route messages to nodes that are closer to the destination node in the mobility pattern space. Node locations are required to construct mobility patterns.

Musolesi et al. [20] propose an adaptive routing protocol for intermittently connected mobile ad hoc networks. They use a Kalman filter to compute the probability that a node delivers messages. This protocol assumes group mobility and cloud connectivity; that is, nodes move as a group, and among this group of nodes a contemporaneous end-to-end connection exists for every pair of nodes. When two nodes are in the same connected cloud, destination-sequenced distance-vector (DSDV) [21] routing is used.

Erasure coding [11,25] explores coding algorithms to reduce message replicas. The source node replicates a message m times, then uses a coding scheme to encode them in one big message. After replicas are encoded, the source divides the big message into k blocks of the same size and transmits a block to each of the first k encountered nodes. If m of the blocks are received at the destination, the message can be restored, where $m < k$. In a uniformly distributed mobility scenario, the delivery probability increases because the probability that the destination node meets m relays is greater than it meets k relays, given $m < k$.

Island Hopping [22] considers a network topology that consists of nodes of a set of connected subgraphs, or *connectivity islands*, and a set of edges representing possible node movements between those islands. Nodes learn the entire graph by exchanging their views of the network based on prior node movement.

We did not implement and evaluate these routing protocols because either they require unrealistic future information [12], controllable mobile nodes [18], or domain-specific information (location information) [16,17]; they assume certain mobility patterns [20,22]; or they present orthogonal approaches [11,25] to other routing protocols.

A recent study by Conan, Leguay, and Friedman [5] is similar to our work. They use the intercontact time to calculate the expected delivery time. Their relay strategy is based on the minimum expected delivery time. The difference from our work is that our relay strategy is based on the maximization of the contact probability within a given time interval. They also used the Dartmouth data, as well as two other data sets for their simulation. They constructed a subset of the Dartmouth data set, in which all users appear on the network every day. Therefore, their Dartmouth data simulation results have higher delivery ratios and shorter deliver delays than ours.

More recently, Yuan et al. [27] propose Predict and Relay, which uses a time-homogeneous semi-Markov process model to predict the future contacts of two specified nodes at a specified time. With this model, a node estimates the future contacts of its neighbors and the destination, and then selects a proper neighbor as the next hop to forward the message. A synthetic mobility model is used for their simulation. Nodes move among a set of landmark locations with a given probability. It is difficult to compare our work with theirs because they use a different mobility model.

1.1.6 Summary

We simulated and evaluated several routing protocols in opportunistic networks.

We propose a prediction-based routing protocol for opportunistic networks. We evaluate the performance of our protocol using realistic contact traces and compare to five existing routing protocols: three simple protocols and two other prediction-based routing protocols.

Our simulation results show that direct delivery had the lowest delivery ratio, the fewest data transmissions, and no meta-data transmission or data duplication.

Direct delivery is suitable for devices that require extremely low power consumption. The random protocol increased the chance of delivery for messages otherwise stuck at some low-mobility nodes. Epidemic routing delivered the most messages. The excessive transmissions and data duplication, however, consume more resources than portable devices may be able to provide.

Direct-delivery, random, and epidemic routing protocols are not practical for real deployment of opportunistic networks because they either had an extremely low delivery ratio or had an extremely high resource consumption. The prediction-based routing protocols had a delivery ratio more than 10 times better than that for direct-delivery and random routing, and they had fewer transmissions and used less storage than epidemic routing. They also had fewer data duplications than epidemic routing.

All the prediction-based routing protocols that we evaluated had similar performance. Our method had a slightly higher delivery ratio but more transmissions and higher storage usage. There are many parameters for prediction-based routing protocols, however, and different parameters may produce different results. Indeed, there is an opportunity for some adaptation; for example, high-priority messages may be given higher transfer and replication probabilities to increase the chance of delivery and reduce the delay, or a node with infrequent contact may choose to raise its transfer probability.

We must note that our evaluations were based solely on the Dartmouth traces. Our findings and conclusions may or may not apply to other network environments, because users in different network environments may have distinct mobility patterns and thus distinct handoff patterns. We also note that our traces were not motion records, our contact model was based on association records, and messages were synthetically generated. We believe, however, that our traces are a good match for the evaluation of handoff predictions and bandwidth reservations. A more precise evaluation may need motion traces—that is, the physical location of users and models to determine users' connectivity.

References

[1] John Burgess, Brian Gallagher, David Jensen, and Brian Neil Levine. MaxProp: Routing for vehicle-based disruption-tolerant networks. In *Proceedings of the IEEE International Conference on Computer Communications (INFOCOM)*, April 2006.

[2] Tracy Camp, Jeff Boleng, and Vanessa Davies. A survey of mobility models for ad hoc network research. *Wireless Communication & Mobile Computing (WCMC): Special Issue on Mobile Ad Hoc Networking: Research, Trends and Applications*, 2(5):483–502, 2002.

[3] Augustin Chaintreau, Pan Hui, Jon Crowcroft, Christophe Diot, Richard Gass, and James Scott. Impact of human mobility on the design of opportunistic forwarding algorithms. In *Proceedings of the IEEE International Conference on Computer Communications (INFOCOM)*, April 2006.

[4] Mooi Choo Chuah, Liang Cheng, and Brian D. Davison. Enhanced disruption and fault tolerant network architecture for bundle delivery (EDIFY). In *Proceedings of IEEE Global Telecommunications Conference (GLOBECOM)*, 2005.

[5] Vania Conan, Jeremie Leguay, and Timur Friedman. Fixed point opportunistic routing in delay tolerant networks. *IEEE Journal on Selected Areas in Communications (JSAC)*, 26(5):773–782, June 2008.

[6] Kevin Fall. A delay-tolerant network architecture for challenged Internets. In *Proceedings of the Conference on Applications, Technologies, Architectures, and Protocols for Computer Communications (ACM SIGCOMM)*, pp. 27–34, August 2003.

[7] Khaled A. Harras and Kevin C. Almeroth. Inter-regional messenger scheduling in delay tolerant mobile networks. In *Proceedings of IEEE International Symposium on a World of Wireless Mobile and Multimedia Networks (WoWMoM)*, pp. 93–102, June 2006.

[8] Tristan Henderson, David Kotz, and Ilya Abyzov. The changing usage of a mature campus-wide wireless network. In *Proceedings of the Annual International Conference on Mobile Computing and Networking (ACM Mobicom)*, pp. 187–201. ACM Press, September 2004.

[9] Pan Hui, Augustin Chaintreau, James Scott, Richard Gass, Jon Crowcroft, and Christophe Diot. Pocket switched networks and human mobility in conference environments. In *Proceedings of ACM SIGCOMM Workshop on Delay Tolerant Networking and Related Networks (WDTN)*, pp. 244–251, August 2005.

[10] Ravi Jain, Dan Lelescu, and Mahadevan Balakrishnan. Model T: An empirical model for user registration patterns in a campus wireless LAN. In *Proceedings of the Annual International Conference on Mobile Computing and Networking (ACM Mobicom)*, pp. 170–184, 2005.

[11] Sushant Jain, Mike Demmer, Rabin Patra, and Kevin Fall. Using redundancy to cope with failures in a delay tolerant network. In *Proceedings of the Conference on Applications, Technologies, Architectures, and Protocols for Computer Communications (ACM SIGCOMM)*, pp. 109–120, August 2005.

[12] Sushant Jain, Kevin Fall, and Rabin Patra. Routing in a delay tolerant network. In *Proceedings of the Conference on Applications, Technologies, Architectures, and Protocols for Computer Communications (ACM SIGCOMM)*, pp. 145–158, August 2004.

[13] Philo Juang, Hidekazu Oki, Yong Wang, Margaret Martonosi, Li-Shiuan Peh, and Daniel Rubenstein. Energy-efficient computing for wildlife tracking: Design tradeoffs and early experiences with ZebraNet. In *The Tenth International Conference on Architectural Support for Programming Languages and Operating Systems*, pp. 96–107, October 2002.

[14] David Kotz and Kobby Essien. Analysis of a campus-wide wireless network. *Wireless Networks*, 11:115–133, 2005.

[15] David Kotz, Tristan Henderson, and Ilya Abyzov. CRAWDAD data set dart-mouth/campus. http://crawdad. cs. dartmouth.edu/dartmouth/campus, December 2004.

[16] Jason LeBrun, Chen-Nee Chuah, Dipak Ghosal, and Michael Zhang. Knowledge-based opportunistic forwarding in vehicular wireless ad hoc networks. In *Proceedings of IEEE Vehicular Technology Conference (VTC)*, pp. 2289–2293, May 2005.

[17] Jeremie Leguay, Timur Friedman, and Vania Conan. Evaluating mobility pattern space routing for DTNs. In *Proceedings of the IEEE International Conference on Computer Communications (INFOCOM)*, April 2006.

[18] Qun Li and Daniela Rus. Communication in disconnected ad-hoc networks using message relay. *Journal of Parallel and Distributed Computing*, 63(1):75–86, January 2003.

[19] Anders Lindgren, Avri Doria, and Olov Schelen. Probabilistic routing in intermittently connected networks. In *Workshop on Service Assurance with Partial and Intermittent Resources (SAPIR)*, pp. 239–254, 2004.

[20] Mirco Musolesi, Stephen Hailes, and Cecilia Mascolo. Adaptive routing for intermittently connected mobile ad hoc networks. In *Proceedings of IEEE International Symposium on a World of Wireless Mobile and Multimedia Networks (WoWMoM)*, pp. 183–189, June 2005. Extended version.

[21] C. E. Perkins and P. Bhagwat. Highly dynamic destination-sequenced distance-vector routing (DSDV) for mobile computers. *Computer Communication Review*, pp. 234–244, October 1994.

[22] Natasa Sarafijanovic-Djukic, Michal Piorkowski, and Matthias Grossglauser. Island hopping: Efficient mobility-assisted forwarding in partitioned networks. In *Proceedings of IEEE Communications Society Conference on Sensor, Mesh, and Ad Hoc Communications and Networks (SECON)*, September 2006.

[23] Jing Su, Ashvin Goel, and Eyal de Lara. An empirical evaluation of the student-net delay tolerant network. In *Proceedings of the International Conference on Mobile and Ubiquitous Systems (MobiQuitous)*, July 2006.

[24] Amin Vahdat and David Becker. Epidemic routing for partially-connected ad hoc networks. Technical Report CS-2000-06, Duke University, July 2000.

[25] Yong Wang, Sushant Jain, Margaret Martonosi, and Kevin Fall. Erasure-coding based routing for opportunistic networks. In *Proceedings of ACM SIGCOMM Workshop on Delay Tolerant Networking and Related Networks (WDTN)*, pp. 229–236, August 2005.

[26] Yu Wang and Hongyi Wu. DFT-MSN: The delay fault tolerant mobile sensor network for pervasive information gathering. In *Proceedings of the IEEE International Conference on Computer Communications (INFOCOM)*, April 2006.

[27] Quan Yuan, Ionut Cardei, and Jie Wu. Predict and relay: An efficient routing in disruption-tolerant networks. In *Proceedings of ACM International Symposium on Mobile Ad Hoc Networking and Computing (MobiHoc)*, May 2009.

Chapter 2

State of the Art in Modeling Opportunistic Networks

Thabotharan Kathiravelu and Arnold Pears

Uppsala University

Contents

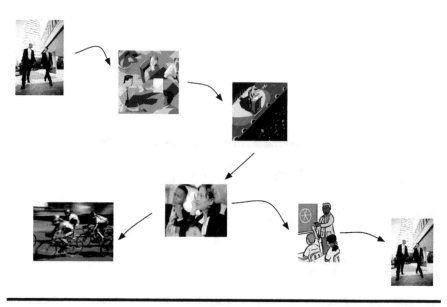

Figure 2.1   A typical example of formation of opportunistic contacts and the opportunistic forwarding of content of interest.

2.1 Introduction

Opportunistic networking explores the data transmission potential of small mobile devices, such as mobile phones and Personal Digital Assistants (PDAs). Small handheld wireless devices carried by humans form opportunistic networks and can utilize intermittent contact with other devices to exchange data [10,28]. Potential opportunistic networks arise as device users congregate and disperse [38]. For example, when people travel in a subway train or in a commuter bus they are in close proximity to other passengers and their mobile devices can contact each other to transfer information—for instance, an MP3 file. Figure 2.1 shows the typical formation of an opportunistic networking environment and how the content of interest is forwarded opportunistically.

Current research in opportunistic networks ranges from opportunistic forwarding strategies, through modeling device contacts, to identifying appropriate use cases and evaluation scenarios, and developing theoretical and mathematical models. Performance modeling and analysis are of both practical and theoretical importance. Since activities in opportunistic networking are still in their early stages, performance modeling is an essential ingredient of opportunistic networking research. Developing better methods for measuring system performance is also of great importance. Performance modeling helps to identify key challenges in both design and research methodologies, as well as to gain insight into the behavior of opportunistic networking systems.

Simulation-based studies are widely used, and one of the main objectives of simulations is to provide support for new ideas. Simulation experiments explore the characteristics of opportunistic systems in a wide range of potential deployment scenarios and help developers to evaluate systems prior to deployment. Researchers have been actively studying device contact patterns, analyzing their frequency, duration, and predictability in order to evaluate mechanisms for the exchange of content among peers.

Modeling and performance evaluation constitute two major research activities. Modeling device contacts is one direction of research, where new models that mimic pairwise device contacts in opportunistic networking environments have been proposed [34,45]. Validating the performance of opportunistic protocols and applications is another direction of research activity. This involves developing, testing, and validating protocols and applications and identifying factors that influence opportunistic content distribution.

Intermittent connectivity between mobile nodes and the behavioral patterns of users make modeling and measuring the performance of opportunistic networks a challenging job. High-fidelity models are needed in order to establish the feasibility of data forwarding protocols and content distribution schemes, to predict performance boundaries for opportunistic applications as well as to characterize the power and memory behavior of the network, transport, and application protocols. A significant challenge in the simulation of opportunistic networks is modeling the underlying pairwise contacts between nodes. Ideally, simulations should closely emulate the behavior of "real" networks.

In the remainder of this chapter we provide an overview of the state of the art in modeling opportunistic network connection structures and pairwise connectivity. We summarize work in analyzing pairwise contacts in collected opportunistic connectivity traces and their findings. We then describe the role of mobility models in modeling contacts in the Mobile Ad hoc Networking research community and argue that new models are needed for opportunistic network research. We introduce connectivity models as one feasible approach to modeling contact in opportunistic networks. The scope of the approach is illustrated using a case study that explores the impact of variation in connectivity properties of the underlying network on an application that distributes content to participants in the network.

2.2 Background

2.2.1 Trace Collection and Analysis

Collecting interdevice contact traces has gained much attention in Mobile Ad hoc Networks (MANET) and opportunistic networking research [10,41]. The main purpose of trace collection is to characterize interdevice contact patterns. Such characterizations are then used to validate new protocols and applications. Many researchers are convinced that collected traces will reveal many interesting facts

Figure 2.2 Small mobile devices with Bluetooth wireless connectivity are being prepared for trace collection in a conference environment in December 2007.

about opportunistic contacts among devices [10,39,41]. Figure 2.2 shows the preparation activities prior to opportunistic contact trace collection in an academic conference in December 2007.

In 2004, Su et al. began investigating the feasibility of opportunistic data forwarding using human connectivity traces that were collected using PDAs [58]. Pioneering work in collecting true opportunistic contact traces and analyzing them for their underlying properties and the feasibility of forwarding using opportunistic contacts was carried out by Chaintreau et al. in 2005 [12]. Chaintreau et al. have analyzed three publicly available traces from the University of California, San Diego (UCSD) [42], Dartmouth University [26], and the University of Toronto [58]. UCSD traces include the client-based logs of the visibility of access points for a three-month period, while the Dartmouth University traces include SNMP* logs from access points for a four-month period. The University of Toronto trace set was collected using 20 Bluetooth-enabled PDAs carried by students for a period of 16 days.

In addition to these traces, Hui et al. have conducted their own experiments with iMote devices and have collected three different contact trace sets [28]. Of these three trace sets, the Intel and Cambridge traces were collected in research lab settings, whereas the iMotes were carried by researchers and doctoral students. The Infocom trace set was collected at the IEEE INFOCOM 2005 conference and the iMotes were carried by the conference participants.

iMotes are a common platform for connectivity trace collection activities. Designed by Intel, they use Bluetooth wireless technology to log their device

---

* Simple Network Management Protocol.

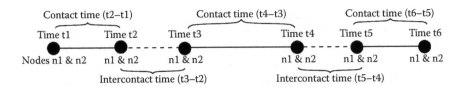

Figure 2.3   Characterization of contact and inter-contact times for two nodes *n*1 **and** *n*2, **which meet intermittently.**

discovery and contact details with other Bluetooth-enabled devices [10,41]. Mobile phones and PDAs are also commonly used in collecting connectivity traces and are a good choice in terms of memory and battery power, but, at the same time, they are expensive to deploy on a larger scale [35].

All these collected traces were analyzed for their interdevice connection properties in order to understand the constraints of opportunistic data transfer, and two new terms—*contact time* and *inter-contact time*—were defined and characterized:

Contact Time is the time interval during which two mobile devices are in each others' communication range and communicate with each other. In experiments based on opportunistic networks, contact times are often recorded and used as a measure for estimating the amount of data that could possibly be transferred. Contact times also are used to determine the pattern of underlying probabilistic distributions over time.

Inter-Contact Time is the time interval measured between two consecutive contacts when two mobile devices are in contact with each other intermittently. The inter-contact time values and their distribution over the period of time are two important pieces of information. The likelihood that a device will be encountered again in a given time period can also be determined or estimated from these values.

In Figure 2.3, we characterize the nature of the contact and inter-contact times of two nodes, *n1* and *n2*, that meet each other intermittently.

Many others followed the path of Chaintreau et al. in collecting traces from different environmental settings and have analyzed the inter-device contact patterns and their distributions. Leguay et al. [41] did an experimental study in the city of Cambridge with 54 Bluetooth-enabled iMote devices that were carried by students or fixed at popular places. Contact data was logged for a period of 13 days. The connectivity information from this experiment was then used in investigating the feasibility of a citywide content distribution network. By categorizing users with different behavioral patterns, they analyzed the effectiveness of this user population in distributing data bundles.*

---

* Data bundle is a term commonly used in delay-tolerant networking [6,9,32] to refer to aggregated data. The same term has also been adopted by opportunistic networking researchers to refer to clusters of data packets [32,41].

LeBrun et al. [39] have explored the feasibility of disseminating information using transit buses on the University of California Davis campus. Each bus is equipped with a Bluetooth content-distribution cache called BlueSpot. Any device with a Bluetooth-enabled interface can connect to these caches and download files while the bus is en route. They predict the provision of services like on-demand downloads of iTunes® music files as a future killer application.

The Reality Mining project [19] at MIT* collected communication proximity, location, and activity information of 100 selected participants over a nine-month period. Each participant was given a mobile phone that had both Bluetooth and cellular network accessibility. This research looked at the mobile device interactions in many directions.

In Table 2.1 we present some opportunistic contact traces that are publicly available at the CRAWDAD† web site. Cambridge1 was collected by Leguay et al. [41], Intel, Cambridge2, and Infocom were collected by Chaintreau et al. [12], and BSpot1, BSpot2, and BSpot3 were collected by LeBrun et al. [39]. The Access Technology mentioned as BT in the table refers to Bluetooth wireless technology [21].

The analysis of the underlying probabilistic distributions of contact and inter-contact times of all collected traces have helped researchers in developing theoretical foundations for opportunistic forwarding algorithms and strategies [11,13]. While trace collection has been, and remains, a popular activity, traces typically have relatively short time periods and small populations, due to the logistic challenges associated with large-scale experiments [29,40,41]. Short logging periods, lack of variation in the device population, and other statistical properties limit the usefulness of many existing traces [60]. Leguay et al. [41] describe the difficulties caused by device failures due to memory overflows, drained battery power, and so on in trace collection processes. Nordström et al. [53] provide insight into the challenges in measuring human mobility and collecting connectivity traces and caution researchers about the limitations inherent in using publicly available connectivity trace sets.

2.2.2 Modeling Contacts

Many research studies in MANET and opportunistic networks are dependent on simulation-based studies to test and validate new hypothesis and ideas. Modeling inter-device contacts is essential in these studies in order to tell the simulation engine how the contacts are made between simulation entities [8,31,37,44]. Researchers have so far been using synthetic mobility models and human contact traces as input for simulation based studies. Recently, there have been attempts to model contacts more accurately by refining mobility models and modeling contacts using the measured properties of human connectivity traces [36,65]. We describe

---

* Massachusetts Institute of Technology.
† Community Resource for Archiving Wireless Data at Dartmouth (CRAWDAD) http://crawdad.cs.dartmouth.edu.

Table 2.1 Properties of Some Publicly Available Opportunistic Connectivity Traces

Device Properties	Cambridge1	Intel	Cambridge2	Infocom	Bspot1	BSpot2	BSpot3
Device	iMote	iMote	iMote	iMote	iMote	iMote	iMote
Access Technology	BT	BT	BT	BT	BT	BT	BT
Duration (days)	13	3	5	3	4	5	6
Granularity (seconds)	120	120	120	120	120	120	120
No. of Internal Devices	54	9	12	41	20	35	33
No. of Contacts	8545	2766	6732	28216	6131	11796	11742
No. of External Devices	11357	118	203	233	512	461	586

mobility models in detail and then look at the recent extensions to these models for intermittently connected mobile networks. We conclude by discussing an alternative approach, *connectivity models*.

2.2.3 Mobility Models

Mobility models generate plausible sequences of inter-device contacts synthetically based on a set of parameters that characterize the environment. Though categorizing any mobility model as realistic is difficult, modeling device mobility is an essential part of any simulation study. Synthetic mobility models are often preferable since they allow the researcher to alter the parameters of the simulation set and reproduce simulation results easily.

The random waypoint (RWP) mobility model is one of the most widely used single-entity random movement methods [4]. In RWP, nodes move in a given direction with a given speed and pause occasionally to change direction or speed. Random walk and random direction are also two widely used single-entity mobility models [8,32,55]. A good survey of synthetic mobility models, excluding a few of the most recent mobility models, can be found in [8].

It has been shown [8] that traditional synthetic mobility models do not model node mobility in a manner that causes the contact and inter-contact time distributions of simulated nodes to decay exponentially over time [5,57,64]. Such a behavior is very different from what has been observed in field tests [10]. Analysis and systematic studies of the stochastic properties of earlier models have identified flaws in modeling, and new mobility models that mimic real-world environments and related movement patterns have been proposed [27,31,36,37,46].

Jardosh et al. propose a mobility model that includes obstacles in the node path in order to mimic mobility in a more realistic environment [31]. In addition to obstacles along the path of node movement, MobiREAL [37] considers the behavioral changes of nodes due to the surroundings, time of the day, and so on, and uses a rule-based model to describe the movement patterns of nodes.

The weighted waypoint mobility model [27] recognizes that people do not choose their waypoints randomly. It models this behavior by defining popular locations in the simulation field and their related popularity *weight* values according to the probability of choosing these locations as the next destination. In this Markov chain–based model, the selection probability of a new destination is determined based on the current location and time.

Mobility models could also be designed based on the observed statistical properties from human mobility traces. The observed statistical properties can then be fed into the mobility model so that the generated interdevice connectivity resembles the same statistical properties [36,65].

Kim et al. [36] describe a method for extracting user mobility tracks from recorded Wi-Fi traces. They have examined the tracks in the Wi-Fi traces and were able to extract information such as speed, pause times, destination transition probabilities, and waypoints between destinations. This information is then used in forming an empirical model that can be used to generate synthetic tracks of movements of nodes [36]. They have validated their model by performance comparisons between trace locations and the locations determined by users carrying both GPS and 802.11 devices.

A new mobility modeling technique has emerged recently based on social network theory [43]. Musolesi and Mascolo devised their mobility model using human social networking concepts and human connectivity traces as the basis for the representation of human movement [46]. Another group's mobility model, also based on the social networking theory that allows nodes to be grouped depending on their social relationships and measured network structures, called the community-based mobility model, was later proposed by Musolesi and Mascolo [44].

Karlsson et al. [32] used random waypoint and random walk mobility models [8] to model node mobility to study the performance of a receiver-driven delay-tolerant content broadcasting environment.

2.2.4 New Generation Models

Recent studies on various networks, such as the Internet and biological networks, have revealed that these networks have fundamentally the same properties as social networks [1,22,48,50]. Empirical studies further reveal that these social networks contain recurring elementary interaction patterns between small groups of nodes that occur substantially more often than would be expected in random networks that are determined by Poisson distributions. These recurring interactions also carry significant information about the network's evolution and functionality. We can also observe similar recurring interactions between nodes in time-aggregated opportunistic field connectivity traces. Please refer to Figure 2.7 for a pictorial view of clusters of node contacts observed in a typical opportunistic field trace set.

The set of clusters of contacts formed by the time-aggregated contacts between pairs of nodes in a connectivity trace set is a new concept in opportunistic networking, even though the concept of clusters is often used in network analysis and graph theory. We view the existence of clusters as follows: Suppose we take a time-aggregated opportunistic connectivity trace set. This aggregated trace set can be represented by a graph, $G(V, E)$, with a set of vertices V representing the nodes and a set of edges, E, representing the aggregated contacts between the node pairs. The edge weights represent the aggregated contact times between pairs of vertices.

A subset $V' \; \forall \; V$ of the vertex set induces a subgraph $G[V'] = (V', E')$ with $E' = \{(u,v) \mid u,v \in V'\}$. A nested decomposition of G by removing edges is a finite sequence $(V_0, V_1, V_2, \dots, V_k)$ of subsets of subgraphs of V such that $V_0 = V$, V_{i+1} is a subgraph of V_i for $i \; \Psi \; k$ and $V_k \neq 0$.

We refer to the subgraphs that arise as a result of subsequent decompositions as the set of *clusters* of nodes. The more edges that are removed, the more clusters are formed, which eventually results in decomposing the graph into individual nodes.

There exist many open source packages, such as iGraph [18] and NetworkX [23], which can be incorporated easily into scripting languages and be used for visualizing network structures and varying natures of clusters of contacts between nodes. Please refer to Figure 2.7 for a sample diagram generated by the iGraph [18] package with the application of the clustering algorithm of Clauset et al. [15]. Many other stand-alone programs and packages are available for the manipulation of networking traces, their properties, community identification, and visualization [52].

Two important terms need to be mentioned here: the *intra-cluster* edges and the *inter-cluster* edges. An edge (*u*, *v*) is an *intra-cluster* edge if both *u* and *v* belong to the same cluster—otherwise it is an *inter-cluster* edge. In Figure 2.7, as mentioned, one can observe the existence of *node clusters, intra-cluster edges*, and *inter-cluster edges* in a typical opportunistic contact trace set after successive removal of edges of less edge weights.

We hypothesize that these cluster structures will be the backbone of the network for these reasons. The edges that remain in the graph are of higher weights and many of them are aggregated values of repeated contacts over the period of time. They represent the recurring nature of contacts that exist between the nodes in a diurnal fashion, and such contacts are capable of forwarding content from one node to another. In addition, ensuring that the content gets distributed to these nodes will ensure that the content gets distributed to the larger user population, since we have seen that these nodes had contacts with other nodes, too, although not as often as they do with the nodes in their own cluster. The ability to model the network at such cluster levels will enable the researcher to model the backbone of the network and the network connections.

Detecting clusters in opportunistic networks allows one to extract information in a more efficient way. In larger networks, identifying the inherent cluster structures narrows down the exploration work to be done. One can use the insight from social sciences to characterize clusters. When used in the analysis of large collaboration networks, clusters reveal the contact patterns, repeated contacts, the informal organization, and the nature of information flow through the whole system [30,49,51].

The algorithm introduced by Newman [49] for identifying communities of contacts is based on the edge betweenness. The *edge betweenness* measures the fraction of all shortest paths passing on a given edge. By removing an edge with high betweenness, one can progressively split the whole network into disconnected components. Further improvements over this algorithm enable researchers to precisely divide a network into a set of clusters [15,51].

As new network features such as resilience and cluster structures emerge, new models that can model graphs with these characteristics and their inherent properties are required. This will enable researchers to model not only contacts in the network but also the substructures (such as clusters of contacts) that are present in the network.

Erdös and Rényi pioneered an approach [20] based on random graphs, where each pair of vertices in the graph is connected by an edge with a given probability. Such models, which are approximated by Poisson processes, are not capable of generating graphs that resemble human dynamics such as social networking and entertainment [2]. Random graphs proposed by Watts and Strogatz [63], and Albert and Barabási [1] exhibit properties like power-law distributions [16] in edge degrees, freedom from scale, and small diameter, and so on, which reflect the complex interaction patterns determined empirically in human dynamics.

Delay-tolerant networks (DTNs) propose a store-carry-and-forward message switching approach as a viable solution to the problem of intermittent connectivity. Devices store messages until they encounter another suitable device, and the devices that receive messages keep on forwarding to other devices along a path that eventually delivers data to the intended recipient. In order to achieve this, the DTN architecture introduces a protocol layer called the *bundle layer* on top of the region-specific lower layers of the protocol stack. This bundle layer handles aggregating data into bundles and delivering them across multiple regions [9].

Characterizing interdevice contacts in delay-tolerant networking would benefit the opportunistic networking community greatly. In a recent work, Conan et al. [17] looked at three reference DTN contact data sets and, based on the statistical analysis on these data sets for their pair-wise inter-contact times, have characterized the heterogeneity in contact and inter-contact times. Conan et al. found that the distributions of inter-contact times are well modeled by log-normal curves for a large number of node pairs.

The node movement patterns in a simulation's coordinate space indirectly result in patterns of pairwise node contacts of varying duration. Connectivity models generate pairwise contacts in the simulation of varying durations using parameters drawn from field traces, obviating the use of a mobility model. The motivation behind this approach is that mobility models are a high-overhead approach to generating pairwise node connectivity data. Therefore, why should the temporal patterns of connectivity not be modeled directly, based on estimated parameters? Generating connectivity information directly relieves the designer of the complexity associated with trying to develop an appropriate model and define its parameters. Conducting experiments with varying numbers of user populations and device connectivity properties, and the comfort of reproducing traces, rather than conducting real field traces, are other advantages of this approach.

2.2.5 Simulation Tools for Modeling and Analysis

Since opportunistic networking is considered as a subclass of MANET research, it has adopted the simulation and other modeling tools used by the MANET community for testing and validating applications and systems. As in MANET research studies, the ease of setup, repeatability of experiments, and the ease of variation in the experimental parameters make simulation-based studies the first choice of opportunistic networking researchers in support of new hypotheses and ideas.

Kathiravelu and Pears [33] used the Jist/SWANS (Java in Simulation Time / Scalable Wireless Ad hoc Network Simulator) simulator, which was developed by Rimon Barr [3], to model and a scenario-based study of an opportunistic networking environment in order to characterize the contact patterns between mobile nodes. In Jist/SWANS, the simulation code written in Java is compiled into a set of class files as usual and then rewritten by the bytecode rewriter into simulation code. The rewritten code is then executed by the simulation engine. Further extensions, such as run time visualization, newer mobility model implementation, and so on, have also been added to Jist/SWANS simulator [14] by researchers. SWANS, which is built on top the JiST simulator, helps researchers to simulate different wireless contact scenarios with different mobility models, such as random waypoint, random walk, and teleport mobility models [33]. The layered architecture of SWANS enables researchers to further extend it according to their needs [33].

Helgason and Jonsson [25] have described their work in customizing the OMNeT++ [61] simulator to model and simulate opportunistic network contacts. They employed two different approaches to simulating opportunistic networking. In the first approach, which is called the *mobility-driven method*, mobility patterns of nodes are specified in a mobility trace Ðle. In the second approach, which is called the *contact-driven method*, the time of node contacts and their durations are specified in a contact trace file. Musolesi and Mascolo [47] also used the OMNeT++ simulator to validate the performance of a new protocol called CAR (Context-Aware Routing) for DTNs using the community-based mobility model [44,45].

In addition, many researchers develop their own simulation engines to carry out experimental studies in opportunistic networks [24,41]. Since very little information is available about these simulation tools, it is not possible to describe them in detail here.

Haggle testbed is a networking architecture designed to cater to the needs of mobile users by providing seamless network connectivity and application functionality in highly mobile environments [56,59]. Haggle separates the application logic from transport binding so that the applications can still be fully functional in highly dynamic mobile environments [59]. Haggle takes care of data handling and communications by adapting to the best available networking environment at that time based on user preferences [59].

2.2.6 A Connectivity Model

In opportunistic networking environments, the aim is to optimize the connection opportunities, and any model that attempts to represent such an environment should consider the factors that affect the most important features of real traces: the *inter-contact time* and the *contact time* [10]. Previous analysis of opportunistic contact traces indicates the significance of these two values, and any synthetic generation process should be capable of reproducing the distributions of these two values.

We propose that connectivity models should be based on the analysis of connectivity patterns in real networking environments. Analysis of these traces for their stochastic properties allows us to derive probability distributions to model internode relationships (the so-called *small world* properties of social networks) as well as the probability of a relationship resulting in two nodes being connected at a given point in time.

Our connectivity model is constructed in two phases:

(i) We determine the social connectedness of the node set.
(ii) After the device relationship network has been determined, the frequency of connection events between two related nodes results in a connectivity trace with the desired properties.

We define a connectivity model M,

$$M = (K, P_R, P_C, N)$$

where

N is the number of nodes in the simulation.

K is a set of clusters $[K_1 .. K_k]$, for a network with k clusters.

$P_R(A, B)$, is the relationship function defining the probability that any two devices A and B are related.

$P_C(e)$, is a connectivity function defining the likelihood that two nodes related by an edge e establish a connection. Here $e \in E$, the set of edges determined by applying the relationship function P_R to all pairs of nodes in the model.

We define P_R, in terms of λ, the intra-cluster relationship distribution, and ϕ, the inter-cluster relationship distribution.

$$P_R(A, B) = \begin{cases} \lambda & \text{where } A, B \in K_i \\ \phi & \text{where } A \in K_i \text{ and } B \in K_j \text{ and } i \neq j \end{cases} \qquad (2.1)$$

From Equation (2.1), $P_R(A, B)$ is used to compute a set of edges E that defines a graph representing the relationships between simulation nodes. If the parameters are chosen correctly, the graph obtained by applying Equation (2.1) to all node pairs will have a cluster and internode edge density to the weighted aggregated graph of the original field trace. See Nykvist and Phanse [54] for an example of such a connection graph.

2.2.7 Application of the Connectivity Model and a Content Distribution Scenario–Based Study

The scenario uses parameters gleaned from the Cambridge1 iMote connectivity traces, published in [41]. It is one of a collection of connectivity traces available from CRAWDAD and is a truly opportunistic contact trace set with a comparatively large number of contacts and devices. From the Cambridge1 data we extracted

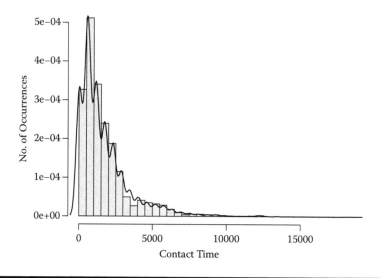

Figure 2.4   Contact time distributions of Cambridge1 [41] trace.

contact time and inter-contact time distributions for iMote devices. We have estimated the *Scale* and the *Shape* parameters for the Gamma distribution visible in the probability density plot of the device contact times. Please refer to Figure 2.4 for the probability distribution plot of the contact times of the Cambridge1 trace set. As has been observed in previous literature, the inter-contact times of the Cambridge1 trace follow a power-law distribution [16] over a 24-hour period. Inter-contact time data for the model is produced using a power-law generator function with coefficient α. The power-law coefficient α was estimated using the maximum likelihood estimation method [16].

We run our simulator using the estimated parameters for the synthetic traces generation, which in turn runs the connectivity model *M*, to generate traces for an equivalent period of time with an equal number of nodes. Generated traces are collected and the contact and inter-contact time values are extracted. First, we compare the generated synthetic trace for its inter-contact time distribution with the inter-contact time distribution of the Cambridge1 trace [41]. The complementary cumulative distribution plot of inter-contact time of both trace sets is presented in Figure 2.5.

We use a scenario similar to the scenario described in [41], which proposes a citywide content distribution architecture. As shown in Figure 2.6, our content distribution architecture is assumed to consist of a couple of short-range wireless devices called *content sources* and other mobile devices with wireless connectivity. Any mobile device that is in direct contact with a content source can opportunistically receive the content. It is further assumed that content sources generate new content on a daily basis and retain the content until the new content for the following day is generated.

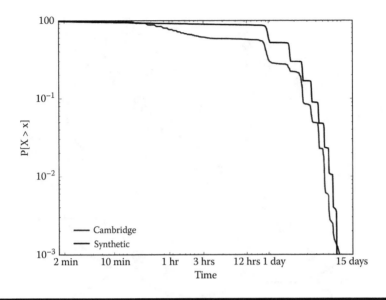

Figure 2.5 **Inter-contact time distribution of Cambridge1 [41] and a synthetic trace set.**

In order to evaluate the content distribution feasibility, we assume the following two distribution schemes:

∎ *Selfish:* In this scheme mobile devices receive the content directly from the content sources and never exchange it with other mobile nodes.
∎ *Cooperating:* In this scheme mobile nodes receive the content directly from the content sources and are ready to share the content with any other mobile nodes that have not received the content yet.

The *selfish* distribution scheme characterizes the feasibility of simple content distribution to small devices. This could be interesting for content push schemes. The *cooperating* content distribution is more advanced compared to the selfish scheme, and it characterizes the feasibility of a true opportunistic content distribution scenario where content is not only injected opportunistically to nodes from content sources, but also is shared opportunistically between nodes that have not yet received it.

We use the following two metrics to measure the performance of the above distribution schemes:

∎ *Delivery Ratio:* This value determines the percentage of the target group that received the content given the time period.
∎ *Average Delay:* The average delay from the time the content was released from the content sources until it is received by target nodes.

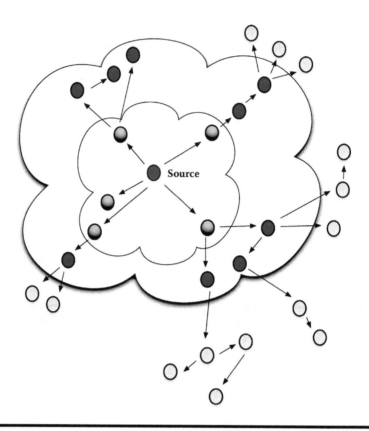

Figure 2.6   A scenario of a citywide content distribution scheme. Nodes at different levels of sharing are drawn in different shades.

In order to evaluate the two above-mentioned distribution schemes and to identify contacts that can be useful for any data transfer, we need to have a threshold value for contact times. Wang et al. [62] did an experimental study of Bluetooth contact initiation times, and from the reported results we assume 2 minutes or 120 seconds as the minimum time value needed for any useful text-based data transfer. We present the performance evaluation results for the two different distribution schemes in the considered scenario in Table 2.2. This table presents the performance results of our candidate field trace Cambridge1 and five different sets of synthetic traces generated by our connectivity model with the same measured parameters.

Though the comparison of the inter-contact time distributions of the Cambridge1 and the synthetic trace set show a close match, the scenario-based performance test results do not show a very close match. In order to analyze the significant differences that were present in the scenario-based evaluations, we wanted to look at the spread of the contacts made throughout the day on an hourly basis. The plot in Figure 2.8

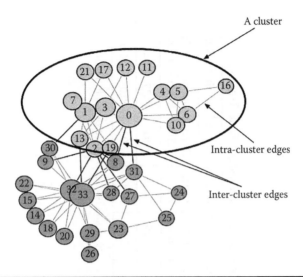

Figure 2.7 **Inter-cluster and intra-cluster edges in clusters of contacts. The clustered graph was generated using the iGraph package [18] by the application of the algorithm proposed by Clauset et al. [15].**

reveals the reason for the differences in the results. A clear observation from the two data sets is that many contacts in the Cambridge1 trace were observed during the later hours of a 24-hour period, whereas in the synthetic trace set, contacts are modeled evenly throughout the 24-hour period and the number of contacts start to decay in the later parts of the 24-hour period as an artifact of the simulation. Therefore, by looking at Figure 2.8, we can make an observation that our model captures the behavior of field traces when we look at the trace set as a whole, but fails to capture the pairwise contact properties in individual cases. Therefore, when we look at the pairwise contacts for the evaluation of the scenario-based case studies, we could not get the expected results. The overall effect is supported by the complementary cumulative distribution plot of inter-contact time distributions, as shown in Figure 2.5, and the probability distribution plot of contact time distributions, as shown in Figure 2.4 and Figure 2.9, where both trace sets show a similar behavior. Hence, from the observations and comparisons of empirical test results, we conclude that our connectivity model is well suited to model network properties at a higher level, and in the future it could be used for such purposes.

2.2.8 Related Work in Connectivity Modeling

Connectivity modeling is an active and open area of research. Work on formulating temporal connectivity models is emerging. Nykvist and Phanse initially proposed connectivity modeling for opportunistic networks [54]. In this work they argue for developing temporal connectivity models directly using the underlying properties

Table 2.2 Comparison of the Application Performance Results of Synthetic Traces and Cambridge1 Field Traces

Method –	Cambridge Calculated		Synthetic Run 1		Synthetic Run 2		Synthetic Run 3		Synthetic Run 4		Synthetic Run 5	
	Delay Ratio	Delay Hours	Delay Ratio	Delay Hours	Delay Ratio	Delay Hours	Delay Ratio	Delay Hours	Delay Ratio	Delay Hours	Delay Ratio	Delay Hours
Selfish	20.00	7.46	10.55	9.96	17.22	9.72	13.33	10.29	12.22	8.91	16.11	12.23
Cooperating	87.77	6.17	86.66	6.64	99.44	7.19	100.0	5.89	99.44	7.03	100.0	6.01

Note: Time threshold = 120 seconds. *Delay Ratio* stands for the delivery ration and *Delay Hours* is the average delay in hours.

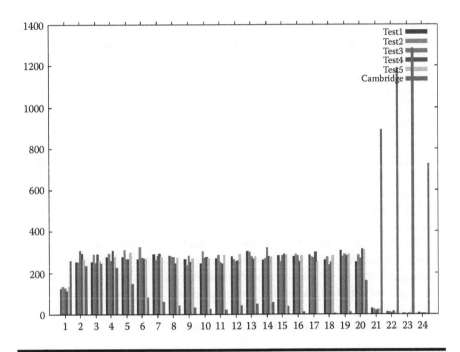

Figure 2.8 **A comparative look at hourly contacts for the Cambridge1 [41] trace set and five different sets of synthetic traces. The horizontal axis shows the hour of the day and the vertical axis shows the number of contacts made.**

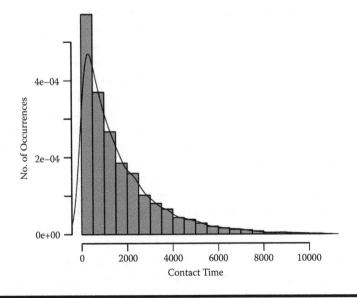

Figure 2.9 **Contact time distribution of a generated synthetic trace set.**

of real movement traces as an alternative to attempting to parameterize mobility features of the traces in order to generate synthetic mobility models.

Calegary et al. have proposed a connectivity trace generator [7]. In this model, the distributions of the contact and inter-contact time are extracted by the connectivity generator and are then used to produce the patterns of collocations of users. The potential connectivity graph for any pair of nodes will be established depending on the distributions of the contact and inter-contact times and the node degree of each node. The connectivity graph is used as a basis for a time-varying graph of instant connectivity for each instance t. In these time-varying graphs, a link between any pair of nodes is either active, if the two hosts are collocated, or inactive, if the two nodes are determined as not collocated. Each link is activated and deactivated according to the distributions of the contact and inter-contact durations.

Chaintreau et al., in analyzing the small diameter characteristic of the opportunistic forwarding paths, have modeled the opportunistic mobile networks as a temporal network in which the network is assumed as a graph with a static set of nodes, and the set of edges between these nodes may change with time [13]. This type of modeling enabled the researcher to establish the existence of the small-world phenomenon in the opportunistic networks.

2.2.9 Open Problems and Challenges

Opportunistic networking, and modeling contacts in opportunistic networks, faces several challenges. There are many research questions to pursue; we now outline some important priority areas in opportunistic networking.

The number of connectivity traces available for the scientific community is limited. Large scale trace sets with varying sets of parameters are needed to cover a broader range of potential opportunistic networking scenarios. Almost all of the available trace sets have been collected in environments with very similar characteristics, typically academic conferences. This creates a problem when evaluating or validating opportunistic network services in other domains; for example, rescue, command and control, online spontaneous gaming, and so on. Collecting large-scale field trace sets is a tedious job and is dependent on many factors. Battery life of small devices, the available memory capacity in devices for logging contacts, corrupted MAC addresses, power-saving setup in devices, and conflict between devices pose severe challenges to researchers collecting large-scale trace sets. Better coordination and collaboration between research groups worldwide, as well as open (well-documented) reference traces are necessary.

Modeling contacts is an active area of research in opportunistic networking. Modeling of contacts mainly depends on the mobility patterns of the socially related parties in considered scenarios. Identifying a candidate scenario that can be mapped to many environments is a key research area. Even if it is difficult to come up with such a single representative scenario, it should be possible to devise approaches to model and develop scenarios and define their properties.

Another interesting challenge is what we are interested in modeling. Connectivity models presented in this chapter do not model the underlying radio properties or the mobility property of concerned nodes. For performance studies, the researcher has to clearly define the impact of different types of abstraction on the validity of the results. Some abstractions may not be suitable in all situations.

We have introduced connectivity models in this chapter and have also discussed mobility models. Connectivity models emphasize a high-level stochastic view of pairwise device connectivity, abstracting away device motion in a physical environment. Mobility models provide information about factors that affect the successful communication between devices and allow physical effects, such as radio propagation, to be taken into account. Though connectivity models and mobility models represent and model different properties of the same set of nodes and scenario, the end result is to facilitate performance testing of systems and applications on different scales and levels of geographical and physical layer fidelity. Integrating them to provide multi-level holistic modeling of opportunistic systems is an interesting problem.

Another open problem of interest is to identify the conditions under which a given opportunistic networking system provides acceptable performance. Identifying parameter variations that lead a system from a functioning state to breakdown is of paramount importance. This will enable system developers to design delay-tolerant content distribution applications with expected rates of failure. In addition, verifying if these conditions are really met in real-time systems is an interesting area of study.

2.3 Chapter Summary

Modeling contacts is an essential part of research in opportunistic networks, as the testing of new services is heavily dependent on simulation-based studies. Modeling contacts enables the researcher to conduct experimental studies to represent various communication environments and measure performance of proposed protocols and services. In this chapter we described early research activities in characterizing opportunistic contacts. Characterizing opportunistic contacts paved the way for definition of two widely discussed properties, contact and inter-contact times. In addition, it also enabled researchers to more clearly understand the nature of the distribution of inter-contact times observed in empirical studies. Characterizing contacts led researchers to consider devising newer models for modeling contacts in opportunistic networks. We provided an overview of research activities in the two distinct types of modeling: mobility models and connectivity models. We discussed the research activities and findings in mobility models in detail, the need to devise more accurate mobility models, and the recent advances in mobility modeling. The need to develop new models based on concepts in modeling contacts that emerge from the social network theory were also summarized.

A new method for modeling contacts, called *connectivity models*, was introduced. We presented a connectivity model and showed its applicability by extracting selected statistical properties from a recent field trace set and applying them to generate a set of synthetic traces. Empirical test results are promising when comparing contact and inter-contact time distributions of the synthetic and original trace sets. A detailed comparative study, based on a content distribution scenario, explored the viability of connectivity models in more detail, and we discussed the reasons for differences in some performance metrics. Finally, we addressed open problems and challenges in modeling opportunistic networks.

References

[1] R. Albert and A. L. Barábasi. Statistical mechanics of complex networks. *Review of Modern Physics,* 74:47–97, 2002.

[2] A. L. Barabási. The origin of bursts and heavy tails in human dynamics. *Nature,* 435(3):207–211, May 2005.

[3] R. Barr, Z. J. Haas, and R. van Renesse. Jist: An efficient approach to simulation using virtual machines. *Software Practice & Experience,* 35(6):539–576, May 2005.

[4] C. Bettsetter, H. Hartenstein, and X. Perez-Costa. Stochastic properties of the random waypoint mobility model. *ACM/Kluwer Wireless Networks, Special Issue on Modeling and Analysis of Mobile Networks,* 10(5), 2004.

[5] C. Bettsetter, G. Resta, and P. Santi. The node distribution of the random way point mobility model for wireless ad hoc networks. *IEEE Transactions on Mobile Computing,* 2(3):257–269, July–September 2003.

[6] S. Burleigh, A. Hooke, L. Torgerson, K. Fall, V. Cerf, B. Durst, K. Scott, and H. Weiss. Delay-tolerant networking: An approach to interplanetary Internet. *IEEE Communications Magazine,* July 2003.

[7] R. Calegari, M. Musolesi, R. Franco, and C. Mascolo. CTG: A connectivity trace generator for testing the performance of opportunistic mobile systems. In *Proceedings of the ESEC and ACM SIGSOFT FSE07,* Dubrovnik, Croatia, September 2007. ACM Press.

[8] T. Camp, J. Boleng, and V. Davies. A survey of mobility models for ad hoc network research. *Wireless Communications & Mobile Computing (WCMC) Special issue on Mobile Ad Hoc Networking Research Trends and Applications,* 2(5):483–502, 2002.

[9] V. Cerf, S. Burgleigh, A. Hooke, L. Togerson, R. Dust, K. Scott, K. Fall, and H. Weiss. Delay tolerant network architecture,draft-irtf-dtnrg-arch-02.txt, July 2004.

[10] A. Chaintreau, P. Hui, J. Crowcroft, C. Diot, R. Gass, and J. Scott. Pocket switched networks: Real-world mobility and its consequences for opportunistic forwarding. Technical report, UCAM-CL-TR-617, University of Cambridge, Computer Lab, February 2005.

[11] A. Chaintreau, P. Hui, J. Crowcroft, C. Diot, R. Gass, and J. Scott. Impact of human mobility on opportunistic forwarding algorithms. *IEEE Transactions on Mobile Computing,* 6(6):606–620, 2007.

[12] A. Chaintreau, P. Hui, J. Crowcroft C. Diot, R. Gass, and J. Scott. Impact of human mobility on the design of opportunistic forwarding algorithms. In *Proceedings of the IEEE INFOCOM 2006,* pp. 1–13, Barcelona, Spain, April 23–29 2006. IEEE explore.

[13] A. Chaintreau, A. Mtibaa, L. Massoulie, and C. Diot. The diameter of opportunistic mobile networks. In *Proceedings of the 3rd International Conference on Emerging Networking Experiments and Technologies (CONEXT '07)*, New York, NY, USA, December 2007. ACM Press.

[14] D. R. Choffnes and F. E. Bustamante. An integrated mobility and traffic model for vehicular wireless networks. In *Proceedings of the 2nd ACM International Workshop on Vehicular Ad Hoc Networks (VANET)*, September 2005.

[15] A. Clauset, M. E. J. Newman, and C. Moore. Finding community structure in very large networks. *Physical Review* E 70, 066111 2004.

[16] A. Clauset, C. R. Shalizi, and M. E. J. Newman. Power-law distributions in empirical data. eprint arXiv:0706.1062, June 2007.

[17] V. Conan, J. Leguay, and T. Friedman. Characterizing pairwise inter-contact patterns in delay tolerant networks. In *Proceedings of Autonomics 2007*, Rome, Italy, October 28–30, 2007. ICST.

[18] G. Csárdi and T. Nepusz. The iGraph software package for complex network research. *InterJournal Complex Systems*, p. 1695, 2006.

[19] N. Eagle and A. Pentland. Reality mining: Sensing complex social systems. *Personal and Ubiquitous Computing*, 10(4), 2006.

[20] P. Erdös and A. Rényi. On random graphs. *Publicationes Methemticae*, 6:290–297, 1959.

[21] E. Ferro and F. Potorti. Bluetooth and Wi-Fi wireless protocols: A survey and a comparison. *IEEE Wireless Communications*, 12(1):12–26, February 2005.

[22] M. Girvan and M. E. J. Newman. Community structure in social and biological networks. *PROC.NATL.ACAD.SCI.USA*, 99:7821–7826, 2002.

[23] A. A. Hagberg, D. A. Schult, and P. J. Swart. Exploring network structure, dynamics, and function using NetworkX. In *Proceedings of the 7th Python in Science Conference (SciPy2008)*, pp. 11–15, Pasadena, CA, USA, August 2008.

[24] A. Heinemann, J. Kangasharju, and M. Muhlhauser. Opportunistic data dissemination using real-world user mobility traces. In *Proceedings of 22nd International Conference on Advanced Information Networking and Applications—Workshops (WAINA 2008)*. IEEE Computer Society, 2008.

[25] Ö. R. Helgason and K. V. Jonsson. Opportunistic networking in OMNeT++. In *In Proceedings of 1st International Workshop on OMNeT++*, Marseille, France, March 2008.

[26] T. Henderson, D. Kotz, and I. Abyzov. The changing usage of a mature campus-wide wireless network. In *Proceedings of the 10th Annual International Conference on Mobile Computing and Networks (Mobicom '04)*, pp. 187–201, 2004.

[27] W.-J. Hsu, K. Merchant, H. Shu, C. Hsu, and A. Helmy. Weighted waypoint mobility model and its impact on ad hoc networks. *Mobile Computing and Communication Review*, January 2005.

[28] P. Hui, A. Chaintreau, J. Scott, R. Gass, J. Crowcroft, and C. Diot. Pocket switched networks and the consequences of human mobility in conference environments. In *Proceedings of the 2005 ACM SIGCOMM First Workshop on Delay-Tolerant Networking and Related Topics (WDTN-05)*, pp. 244–251, Philadelphia, PA, USA, August 2005.

[29] P. Hui and A. Lindgren. Phase transitions of opportunistic communication. In *Proceedings of the Third ACM Workshop on Challenged Networks CHANTS '08*, pp. 73–80, New York, NY, USA, 2008. ACM Press.

[30] P. Hui, E. Yoneki, S.-Y. Chan, and J. Crowcroft. Distributed community detection in delay tolerant networks. In *Proceedings of the MobiArch'07*, Kyoto, Japan, August 2007. ACM Press.

[31] A. Jardosh, E. M. Belding-Royer, K. C. Almeroth, and S. Suri. Towards realistic mobility models for mobile ad hoc networks. In *Proceedings of Mobicom'03*, San Diego, CA, USA, September 2003.

[32] G. Karlsson, V. Lenders, and M. May. Delay-tolerant broadcasting. In *Proceedings of the 2006 SIGCOMM Workshop on Challenged Networks (CHANTS '06)*. ACM Press, 2006.

[33] T. Kathiravelu and A. N. Pears. What & when: Distributing content in opportunistic networks. In *Proceedings of the International Conference on Wireless and Mobile Computing (ICWMC 2006)*, Bucharest, Romania, July 2006.

[34] T. Kathiravelu and A. N. Pears. Reproducing opportunistic connectivity traces using connectivity models. In *Proceedings of the Third Annual CoNEXT Conference*, New York, NY, USA, December 2007. ACM Press.

[35] S. Keshav, Y. Chawathe, M. Chen, Y. Zhang, and A. Wolman. Panel: Cell phones as a research platform. In MobiSys '07: *Proceedings of the 5th International Conference on Mobile Systems, Applications and Services*, New York, NY, USA, 2007. ACM.

[36] M. Kim, D. Kotz, and S. Kim. Extracting a mobility model from real user traces. In *Proceedings of the 25th Annual Joint Conference of the IEEE Computer and Communications Societies (INFOCOM2006)*, Barcelona, Spain, April 2006.

[37] K. Konishi, K. Maeda, K. Sato, A. Yamasaki, H. Yamaguchi, K. Yasumoto, and T. Hi-gashino. Mobireal simulator-evaluating MANET applications in real environments. In *Proceedings of the 13th IEEE International Symposium on Modeling, Analysis, and Simulation of Computer and Telecommunication Systems (MASCOTS)*, 2005.

[38] J. Lawrence and T. Payne. Exploiting familiar strangers: Creating a community content distribution network by co-located individuals. In *Proceedings of the 1st Workshop on Friend of a Friend, Social Networking, and the Semantic Web*, Galway, Ireland, September 2004.

[39] J. LeBrun and C. N. Chuah. Bluetooth-based content distribution stations on public transit systems. In *Proceedings of the ACM Mobishare'06*, Los Angeles, CA, USA, September 25, 2006.

[40] F. Legendre, V. Lenders, M. May, and G. Karlsson. Narrowcasting: An empirical performance evaluation study. In *CHANTS '08: Proceedings of the Third ACM Workshop on Challenged Networks*, pp. 11–18, New York, NY, USA, 2008. ACM Press.

[41] J. Leguay, A. Lindgren, J. Scott, T. Friedman, and J. Crowcroft. Opportunistic content distribution in an urban setting. In *Proceedings of the ACM SIGCOMM 2006 Workshop on Challenged Networks (CHANTS)*, Pisa, Italy, September 2006.

[42] M. McNett and G. M. Voelker. Access and mobility of wireless PDA users. *Mobile Computing Communication Review*, 9(2):40–55, 2005.

[43] M. Musolesi, S. Hailes, and C. Mascolo. An ad hoc mobility model founded on social network theory. In *Proceedings of the 7th ACM/IEEE International Symposium on Modeling, Analysis and Simulation of Wireless and Mobile Systems*, Venice, Italy, October 2004.

[44] M. Musolesi and C. Mascolo. A community-based mobility model for ad hoc network research. In *Proceedings of the 2nd ACM/SIGMOBILE International Workshop on Multi-hop Ad Hoc Networks: From Theory to Reality (REALMAN'06)*, Florence, Italy, May 2006.

[45] M. Musolesi and C. Mascolo. Designing mobility models based on social network theory. *ACM SIGMOBILE Mobile Computing and Communications Review*, 11(3), 2007.

[46] M. Musolesi and C. Mascolo. Designing mobility models based on social network theory. *Mobile Computing and Communications Review*, 11(3):59–70, July 2007.

[47] M. Musolesi and C. Mascolo. Car: Context-aware adaptive routing for delay tolerant mobile networks. *IEEE Transactions on Mobile Computing*, July 2008.

[48] M. E. J. Newman. The structure and function of complex networks. *SIAM Review,* 45(2):167–256, 2003.

[49] M. E. J. Newman. Detecting community structure in networks. *The European Physical Journal B,* 38:321–330, May 2004.

[50] M. E. J. Newman. Fast algorithm for detecting community structure in networks. *Physical Review E*69, 066133, 2004.

[51] M. E. J. Newman and M. Girvan. Finding and evaluating community structure in networks. *Physical Review E*69, 026113, 2004.

[52] A. Noack. Energy models for graph clustering. *Journal of Graph Algorithms and Applications,* 11(2):453–480, 2007.

[53] E. Nordström, C. Diot, R. Gass, and P. Gunningberg. Experiences from measuring human mobility using Bluetooth inquiring devices. In *Proceedings of MobiEval 2007,* San Juan, Puerto Rico, 2007.

[54] J. Nykvist and K. Phanse. Opportunistic wireless access networks. In *Proceedings of the First International Conference on Access Networks,* Athens, Greece, September 2006.

[55] E. M. Royer, P. M. Melliar-Smith, and L. Moser. An analysis of the optimum node density for ad hoc mobile networks. In *Proceedings of the IEEE International Conference on Mobile Communications (ICC),* St. Petersburg, 197376, Russia, 2001.

[56] J. Scott, P. Hui, J. Crowcroft, and C. Diot. Haggle: A networking architecture designed around mobile users. In *Proceedings of the Third Annual IFIP Conference on Wireless On-demand Network Systems and Services (WONS 2006),* Les Menuires, France, January 2006.

[57] G. Sharma and R. R. Mazumdar. Scaling laws for capacity and delay in wireless ad hoc networks with random mobility. In *Proceedings of ICC 2004.* IEEE, 2004.

[58] J. Su, A. Chin, A. Popivanova, A. Goel, and E. de Lara. User mobility for opportunistic ad-hoc networking. In *Proceedings of the 6th IEEE Workshop on Mobile Computing Systems and Applications (WMCSA),* pages 41–50, Los Alamitos, CA, USA, 2004.

[59] J. Su, J. Scott, P. Hui, J. Crowcroft, E. de Lara, C. Diot, A. Goel, M. How Lim, and E. Upton. Haggle: Seamless networking for mobile applications. In *Proceedings of the Ninth International Conference on Ubiquitous Computing (UbiComp 2007),* Innsbruck, Austria, October 2007.

[60] J. Su, J. Scott, P. Hui, E. Upton, M. H. Lim, C. Diot, J. Crowcroft, A. Goel, and E. de Lara. Haggle: Clean-slate networking for mobile devices. Technical Report UCAM-CL-TR-680, University of Cambridge Computer Laboratory, January 2007.

[61] A. Varga. The OMNeT++ discrete event simulation system. In *Proceedings of the European Simulation Multiconference (ESM '01),* Prague, June 2001.

[62] A. I. Wang, M.S. Norum, and C-H. W. Lund. Issues related to development of wireless peer-to-peer games in J2ME. In *Proceedings of the 1st Conference on Entertainment Systems (ENSYS 2006),* Guadeloupe, French Caribbean, February 2006.

[63] D. J. Watts and S. H. Strogatz. Collective dynamics of small-world networks. *Nature,* 393(4):440–442, June 1998.

[64] J. Yoon, M. Liu, and B. Noble. Random waypoint considered harmful. In *Proceedings of the IEEE INFOCOM 2003,* pp. 1312–1321, San Francisco, CA, USA, April 2003.

[65] J. Yoon, B. Noble, M. Liu, and M. Kim. Building realistic mobility models from coarse-grained traces. In *Proceedings of 4th International Conference on Mobile Systems Applications and Services (MobiSys2006),* Uppsala, Sweden, June 2006.

Chapter 3

Credit-Based Cooperation Enforcement Schemes Tailored to Opportunistic Networks

Isaac Woungang

Ryerson University

Mieso K. Denko

University of Guelph

Contents

3.1 Introduction

Due to recent advances in wireless technologies such as 3G, Bluetooth, IrDA, and WiFi, to name a few, a pragmatic evolution of the generic Mobile Ad Hoc Network (MANET) paradigm, referred to as *opportunistic networking*, has emerged. Opportunistic networks are a kind of challenged mobile ad hoc network, where prolonged disconnections, unpredictable and unstable topologies, and partitions can frequently occur. This results in a paradigm shift for the design of network services. Opportunistic networks are quite different from legacy ad hoc networks in that the network is disconnected as a rule rather than as an exception.

In such networks, nodes are typically controlled by rational entities such as people or organizations, and routing is an inherently cooperative activity. Thus, system operation is expected to be critically impaired due to the possible presence of selfish and/or malicious nodes, unless cooperation in routing and packet forwarding is boosted in some way. Thus, existing routing solutions for legacy ad hoc networks might no longer be suitable for opportunistic networks because of the lack of a contemporaneous path between source and destination nodes, as well as a high variation in network conditions, and long feedback delays, to name a few reasons.

Cooperation in the context of routing has been extensively investigated in the framework of peer-to-peer and ad hoc networks, following three broad perspectives:

(1) the effect that cooperation can have on the overall network performance, (2) methods for detecting non-cooperative behaviors, and (3) design of mechanisms to enforce cooperation. Of course, these challenges are interdependent and should be treated as a whole. They lessen the need for identifying incentive protocols and incentive-aware routing schemes that can be applied to opportunistic networks. Consequently, this chapter departs from the study of cooperation in the presence of selfish and/or malicious nodes in legacy ad hoc network environments such as the Internet, mobile ad hoc networks, wireless networks, and peer-to-peer systems in general.

Most existing work on incentive mechanisms in these area networks fall into three categories:

1. *Reputation-based schemes*—These schemes attempt to identify misbehaving nodes and isolate them from the network. In these schemes, a set of trusted nodes are used to detect, assess, and validate the misbehavior of selfish and/or malicious nodes and prohibit them from participating in the network's activities. Typically, nodes are motivated to participate as relays in data forwarding because of their fear that, if detected, they will be punished.
2. *Credit-based schemes*—These schemes assign credits to nodes that forward packets. Typically, credits are taken away from nodes that do not cooperate and credits are given to those that participate in packet forwarding. In this capacity, a dedicated secure hardware or third trusted party is responsible for the management of credits.
3. *TFT-based schemes*—These schemes use a Tit-for-Tat (TFT) strategy to "reward" *well-behaved* nodes and "punish" *badly behaved* nodes. Typically, a node lowers service to its neighbor if it detects bad behavior in the neighbor and fully cooperates with its neighbor if good behavior is acknowledged.

In this chapter, we focus only on credit-based protocols and incentive-aware routing schemes by proposing a qualitative comparison of existing incentive schemes based on predefined incentive scheme characteristics [1,2]. It should be emphasized that this work extends the works in [1,2] in at least two ways. First, our work focuses on more recent incentive protocols and incentive-aware routing schemes. Second, our goal is to identify incentive protocols and incentive-aware routing schemes that are suitable for opportunistic networks. To our knowledge, no such comprehensive and extensive survey exists.

The rest of this chapter is organized as follows. Section 3.2 presents the opportunistic networking paradigm. Section 3.3 sets a context for our study. A set of characteristics of incentive patterns inherited from earlier studies are described. These characteristics are used in Section 3.4 to evaluate the studied credit-based schemes. Finally, Section 3.5 concludes our work.

3.2 Opportunistic Networking Paradigm

Over the years, the need for increasing the benefit of computer resources has grown tremendously, leading to new advances in wireless technologies, with the goal of enabling ubiquitous communication and pervasive computing. Some instances of pervasive computing environments include distributed computing environments, ubiquitous computing environments, mobile and ambient networking environments, and, most recently, opportunistic networking environments.

3.2.1 Types of Opportunistic Networks

The opportunistic networking paradigm has mostly been motivated by the desire to move from a network of devices to which users must adapt to a network that adapts itself to user needs and behaviors. From a practical perspective, it can be viewed as a pragmatic evolution of the generic MANET paradigm.

Opportunistic networks are self-organized mobile wireless networks that opportunistically use all kinds of communication possibilities that wired or wireless devices can offer. In the current research trends, one can distinguish three major types of opportunistic networks [3]:

1. *Opportunistic Communication Networks* (Class 1 Opportunistic Networks): The only goal of these networks is to enable, on the fly, the establishment of connections and communications among previously disconnected nodes. Illustrative studies of such networks can be found in [4,5,6].
2. *Opportunistic Data Dissemination Networks* (Class 1.5 Opportunistic Networks): These networks opportunistically propagate and forward data. Examples of such networks can be found in [7,8,9,10].
3. *Opportunistic Capability Utilization Networks* (OCUNs) (Class 2 Opportunistic Networks): This is a more generic class in the sense that its role encompasses communication and data dissemination and includes such other capabilities as resource management, services discovery, and management, to name a few. An instance of such networks has been studied in [11]. Two distinguished subclasses of this class of networks are:
   a. A novel lower-level pervasive computing paradigm [12], referred to as Oppnets.
   b. Delay-/disruption-tolerant networks (DTNs).

The first class promotes the idea that the network architecture can grow or shrink opportunistically by discovering, analyzing, inviting, or integrating/releasing new available pervasive resources in order to accomplish the network's goal. The second class, also called a network of regional networks, has recently attracted a lot of interest [10,14,15,16,17], due to their suitability for opportunistic networking applications [13]. Examples of DTNs are pocket switched networks, pervasive networks made of users' devices only, socio-aware community networks, wildlife tracking sensor networks, autonomic networks, and so on.

The growth of the aforementioned network architecture and shrinking capabilities is what distinguishes Oppnets from DTNs, OCUNs, and other pervasive and opportunistic computing technologies. This chapter investigates incentive-aware routing schemes and incentive protocols that can be used to stimulate nodes in opportunistic networks to cooperate in the routing process.

3.2.2 Routing in Opportunistic Networks

In legacy Mobile Ad Hoc Networks (MANETs), and more generally in existing popular Internet applications, the routing and data forwarding processes assume the existence of a pre-established contemporaneous end-to-end path between the source and destination nodes (for instance, determined via a secure ad hoc routing protocol based on Dynamic Source Routing (DSR) [18]), characterized by moderate round-trip times and leading to small packet loss probabilities. In this case, incentive schemes for routing are geared toward *route discovery only*. In opportunistic networks, a path between the sender and destination nodes is computed on the fly—i.e., in a dynamic fashion—depending on the availability and willingness of the intermediate node to forward the received data bundle. Typically, a sender may not be able to predict the bundle forwarding path, and intermediate nodes may suffer from frequent loss of connectivity (referred to as *intermittent connectivity*) or long round-trip delays to the sender. In this case, opportunistic network protocols [19], that is, protocols that opportunistically exploit one-hop communication opportunities between nodes to bring data (in a store-carry-forward fashion) closer to potential interested nodes, data-centric networking (as opposed to conventional topology-centric networking), and node mobility, are advocated as key routing features that should be considered when designing incentive protocols for routing in opportunistic networks. This is justified by the fact that, in the these schemes, it is expected that both route establishment and route discovery mechanisms are implemented, as well as strategies for managing the node's storage capacity and cooperation among nodes.

Part of the contribution of this chapter is to describe what the mechanisms are for some of the studied incentive schemes. Of course, these mechanisms rely on their own underlying opportunistic routing approaches. We refer the reader to [9], [13], and [20] for a comprehensive review of routing protocols in opportunistic networks. These mechanisms also depend on what motivated their design as discussed in the next section. It should be recalled that in opportunistic routing, a node overhearing a packet has the capability to participate in its forwarding process in a store-carry-and-forward fashion; that is, the routing consists of independent and local forwarding decisions based on current connectivity and predictions of future connectivity information. Broadcast transmissions are utilized to send information through multiple concurrent relays. This is in contrast to the conventional topology-based routing approach where the next hop of a node is deterministically chosen before the transmission of the data packet occurs.

3.2.3 Motivation for Studying Cooperation Schemes Tailored to Opportunistic Networks

Due to the versatility of the opportunistic networking paradigm, several instances of opportunistic network architectures can be identified [13,21,22]. All these can be viewed as evolutions of the generic MANET paradigm [9], which advocates the transfer of messages from source to destination performed by intermediate nodes in a *store-carry-and-forward* fashion. Thus, the motivations for studying incentive schemes tailored to opportunistic networks naturally inherits from those sustaining the design of existing incentive schemes for mobile wireless ad hoc networks.

A few reasons why nodes in an opportunistic network application could voluntarily decide not to cooperate in the routing or bundle forwarding process have been identified as follows: (1) Battery power and availability of transmission bandwidth are critical values of interest that could be used by nodes to deny forwarding of messages from other nodes. (2) When operating normally in the route discovery and route maintenance phases, a selfish node may be reluctant to cooperate with other nodes if no direct benefit or compensation for its involvement is guaranteed. For instance, it may be reluctant to act as a packet relay to save its resources such as battery life, CPU cycles, and available bandwidth. (3) For the above Oppnets subclass [12], an Oppnet must first obtain a formal agreement with a helper that is willing to sign up and participate in its activities. Such agreement (for instance, based on economical advantages such as micropayments or promises of stipends [23,24]) could affect the helper's decision to join the Oppnet. To cope with these difficulties, several incentive protocols and incentive-aware routing schemes for autonomic networks and wireless mobile ad hoc networks have been proposed and compared based on identified characteristics [2]. The work in [2] was further complemented by the work in [1], where the authors provided a classification of incentive patterns according to specific requirements of the application domain and environment.

This chapter builds on these pioneering works by complementing them with a study of the most recently proposed credit-based incentive protocols (not covered in [1,2]) as well as incentive-aware routing schemes. A comparison of the studied incentive schemes against a set of incentive pattern characteristics inherited from [1,2] is proposed. Incentive schemes that are the most relevant to opportunistic networks are then highlighted. Of course, a context for the study must first be defined.

3.3 Context of Our Study

The characteristics of the studied incentive schemes that will be used for comparison purposes are determined on the basis of some contextual definitions, as follows.

3.3.1 Elementary Cooperation

In opportunistic networks, each node is expected to behave autonomously and to forward its data to other nodes in an opportunistic manner. However, this network functioning cannot always be directly assumed because the provision of a service by one node to another consumes resources, and each node may try to maximize its own utilities in a self-interested fashion. Therefore, cooperation is a primary requirement.

In this chapter, it is assumed that the cooperation between the entities (i.e., resources, nodes, networks, subnetworks, protocols) is to be treated at the *elementary level*, as discussed in [1]. More precisely, when entity A acts on behalf of entity B and offers some services, A is called a *provider* and B is called a *consumer*. In other words, A provides a service that is beneficial to B and might or might not request any compensation from B in return.

3.3.2 Layered Cooperation

We have adopted the node's protocol stack (shown in Figure 3.1) that was proposed in [19] for delay-tolerant networks. In this setting, cooperation is also concerned with each layer of the node's protocol stack. For this reason, we use the conceptual layering model introduced in [2], which determines a way to define what it means to have uncooperative behavior between two nodes, thereby allowing for a classification of interdevice cooperation.

The protocol stack depicted in Figure 3.1 uses a bundle layer [52], which acts as a novel and additional protocol layer between the transport and application layers. For the purpose of allowing the providers to offer application services that can be beneficial to consumers, we have added a *discovery* sublayer to the application layer.

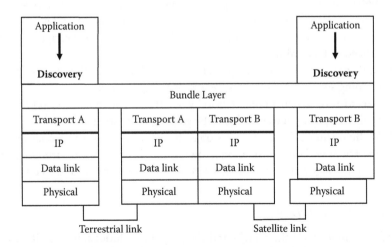

Figure 3.1   A node's protocol stack in an opportunistic network [19].

A detailed description of each layer of the above-mentioned protocol stack can be found in [19].

Most characteristics of incentive patterns and incentive schemes used in this chapter are taken from [1,2]. Thus, we will not describe or justify them here. The notations and definitions that follow are only meant to clarify the above-mentioned characteristics, which we use in Section 3.4 to evaluate the studied incentive schemes.

3.3.3 Notations and Definitions of Basic Terms

In this section, we will give the notations and describe some basic terminologies and comparison attributes used for various comparisons made in this chapter. The notations and definitions include node misbehaviors, attack types, cooperation and transaction processing, incentive schemes, security and trust models, implementation details payment models, and data dissemination.

3.3.3.1 Node Misbehaviors

In ad hoc networks and challenged networks, most routing protocols that have been proposed generally assume that the nodes will fully participate in the data forwarding process. Unfortunately, this behavior is not always guaranteed because nodes may misbehave for selfish or malicious reasons, or a faulty link from the wireless medium may exist. This chapter deals with four types of node misbehaviors:

- **Selfishness:** When a node is reluctant to forward packets destined for other nodes without gain. This category also includes: *individual selfishness*—when only a small subset of selfish nodes do not wish to forward packets to others, and *mass selfishness*—when every node drops a certain percentage of packets instead of forwarding them.
- **Rationality:** When a node only attempts to cheat when the expected benefit of doing so is greater than the benefit of acting honestly.
- **Greediness:** When a node tries to bypass the incentive scheme and cheat for credits by paying fewer credits or by gaining more credits.
- **Maliciousness:** When a node's objective is to interrupt the proper network operations without considering its own gain. This can include dropping packets, injecting traffic, exploiting a weakness in the network, and so on.

3.3.3.2 Attack Types

In order to understand the attacks and the proposed solutions, we classify attacks into several groups (which are not necessarily mutually exclusive). The types of attacks that we deal with in this chapter are summarized in Table 3.1.

Table 3.1 Definitions with Respect to Attack Types

Attack Types	
Attacks by Selfish Nodes	*Meaning*
1: Credit fraudulence (or forgery) attack	When a selfish node cheats by attempting to fabricate a valid credit in order to reward itself for work it did not do or to receive more rewards.
2: Repudiation attack	When the source or destination node denies payment for previous communications realized through intermediate nodes or when the source node colludes with the destination node to deny paying the credits to intermediate nodes for their participation in packet forwarding.
3: Node collusion attack	When several nodes collude together to cheat in order to escape being charged.
4: Free riding attack	When two misbehaving nodes on the forwarding path attempt to exchange packets without paying for them.
5: Individual selfishness attack	When only a small set of selfish nodes do not wish to forward packets for other nodes.
6: Mass selfishness attack	When each node drops a certain percentage of packets instead of forwarding them. For instance, this can be realized in a probabilistic fashion.
7: Nodular tontine attack	When a selfish node attempts to gain extra credits from another node by removing it from the bundle forwarding process.
8: Submission refusal attack	When a node that appears to be the last one in the established forwarding path refuses to collude with the source node to receive fake compensation that it does not merit.

(Continued)

Table 3.1   Definitions with Respect to Attack Types (Continued)

Attack Types	
Attacks by Selfish Nodes	*Meaning*
9: Denial of service attack	When a selfish node in the network attempts to flood the network with useless information or prevents other nodes from accessing services or information.
10: Acknowledgment refusal attack	When a node receives a message but does not acknowledge or report its receipt.
11: Forwarding refusal attack	When a node receives a message (or packet) but deliberately refuses to forward it; or when intermediate nodes that have obtained the credits from source nodes are reluctant to forward packets for the source nodes.
12: False claim attack	When a node did not receive a message but falsely claimed that it did receive it.
13: Protracted vendetta cheating	When a node delivers fewer packets than expected to another node; for instance, due to variation in mobility.
Attacks by Malicious Nodes	*Meaning*
14: Deviating from a defined mechanism	When a malicious node cheats by attempting to deviate from the defined mechanism in order to obtain more credit.
15: Exploiting a weakness of an algorithm	When a set of malicious nodes aim to unbalance the system by exploiting a weakness of the algorithms.
16: Packet forwarding misbehavior	When packet forwarding misbehavior occurs as a malicious activity.
17: Routing misbehavior	When the routing misbehavior occurs as a malicious activity.

3.3.3.3 Cooperation and Transactions

In ad hoc networks, devices must cooperate in order to compensate for the absence of infrastructure. Thus, incentives are indispensable to induce cooperation between devices on specific protocol layers. In this context, we refer to the following definitions [1,2]:

- **Cooperation domain:** Refers to the protocol layer in which interdevice cooperation happens; this could be the application layer, discovery layer, transport layer, network layer, data link layer, or any combination of those layers.
- **Transaction:** This is defined as the relay of a message from a sender to a destination (a case of incentive-aware routing schemes) or as elementary principal–agent cooperation (a case of incentive protocols). In the latter case, one can distinguish between negotiation and processing.
- **Negotiation:** First phase of the elementary level cooperation, intimating that the provider and the consumer have both agreed on the provider's action.
- **Processing:** Second phase of the elementary level cooperation, consisting of a provider's action followed by a kind of remuneration.

3.3.3.4 Incentive Patterns and Incentive Scheme Characteristics

In this chapter, we deal with two types of credit-based incentive protocols:

- **Credit-based incentive-compatible routing protocols:** This type of protocol is denoted as IncCOR.
- **Credit-based incentive protocols:** This type of protocol is denoted as IncProt.

In both cases, an incentive pattern is defined as a stimulus for cooperation that allows an entity to act as a provider for a consumer—that is, to provide a service. An incentive pattern induces several mechanisms that can be used to characterize an incentive scheme. The consumer remunerates the provider by issuing a check or a bearer bill.

Several characteristics of incentive patterns derived from [1] are used in this chapter:

- **Barter trade:** When the incentive pattern advocates an immediate service in return; that is, the consumer must remunerate the provider by simultaneously executing a service in return.
- **Bond based:** When the incentive pattern advocates that the service be deferred in return. The consumer hands over a bond that promises a service (to be fulfilled in the future) in return to the provider. This includes bearer notes, bearer bills, banking, and bank note patterns. The consumer remunerates the provider by issuing a bearer note (a case of a bearer note pattern) or a bearer bill (a case of a bearer bill pattern) or bank notes (a case of a bank note

pattern). For a banking pattern, the incentive scheme assigns a bank account to each entity, followed by a remuneration.

- **Symmetric role:** When, under the incentive pattern's stimulus, an entity acts as a provider for a consumer, with an expectation that the consumer will reward the offered service.
- **Asymmetric role:** When, under the incentive pattern's stimulus, no rule is imposed on the consumer to reward a service operated by the provider on his behalf.
- **Scalability:** This feature of the incentive scheme applies when the incentive pattern applies to a significant number of participating entities.
- **Remuneration type:** Type of remuneration used with respect to the incentive pattern used by the incentive scheme.
- **Remuneration Storage site:** The credits database host used by the incentive scheme.

3.3.3.5 Security and Trust Models

In incentive schemes for ad hoc networks, trust is used either directly or indirectly for enhancing the security of the scheme. The following are some trust definitions used in this chapter:

- **Trust usefulness:** We answer "Yes" if trust is used as an incentive for cooperation and "No" otherwise—that is, if trust is used as a prerequisite for validating the remuneration mechanism.
- **Static trust:** This happens when there is a statement of trust, such as a certificate, a chain of certificates, and so on, using Public Key Infrastructure (PKI) or other cryptography methods.
- **Dynamic trust:** This happens when trust is based on the entities' direct or indirect experience. In this case, a mechanism should be invoked by the incentive scheme to learn from these experiences.
- **Trustee:** Indicates which entities are considered as trusted.
- **Anonymity:** When the participating entities do not have to disclose their identities with respect to trust.
- **Content integrity check:** This applies when the incentive scheme provides a mechanism to check the integrity of the forwarding data. In this case, we indicate which method (if one exists) is used to realize this feature.

3.3.3.6 Implementation and Performance

In terms of implementation and performance of the studied incentive schemes, we use the following vocabulary:

- **Cryptography:** Indicates whether the incentive scheme uses a cryptography infrastructure or not.

- **Tamper resistance:** Indicates which hardware-based technique (if one exists) is used to reinforce the cryptography infrastructure.
- **Validation:** Specifies the validation tool or method used to assess the performance of the incentive scheme.
- **Mobility models:** Specifies the mobility models used (if applicable).
- **Network models:** Specifies which networks are targeted by the incentive scheme.

3.3.3.7 Payment and Rewarding Models

In credit-based incentive schemes, a *credit* represents a kind of stimulus assigned to nodes that have successfully achieved the packets forwarding. Typically, credits are managed through a secure module, hardware, or a third trusted party (TTP) entity. With respect to the payment and rewarding models of the studied credit-based incentive schemes, we employ the following definitions:

- **Transferability:** This happens when a TTP entity is used for credit clearance.
- **Convertibility:** This happens when the conversion of credits from digital to real-world money is facilitated.
- **Tamper-proofness:** In order to guarantee the security of the payment process, any payment method uses a kind of tamper-proofness, realized either through a centralized authority, a TTP, a tamper-proof secure module, or via a method involving public key certificates, identity-based cryptography, or secret-key cryptography, to name a few.
- **Charging model:** This determines the payment or reward model used to assign and reward credits. Typically, a charging rate applies.
- **Charging rate:** This determines how much well-behaved intermediate nodes are paid for their successful participation in the packet forwarding process. All nodes have either the same charging rate or each node has a different forwarding cost (for instance, a payment corresponding to its incurred cost or in the form of equal reciprocal service).
- **Flexible remuneration:** This happens when the assessment of the remuneration depends on the provider's resources and the consumer's needs.
- **Aggregation of remunerations:** This happens when immediate remuneration is infeasible or when a flow control mechanism is invoked by the incentive scheme.

3.3.3.8 Data Dissemination and Path Identification Strategies

In this chapter, we consider the following definitions with respect to data routing:

- **Packet forwarding strategy:** This specifies the approach (if one exists) used to realize the packet forwarding task.
- **Unicast:** This occurs when several copies of the data are sent from the source to each destination.

- **Multicast:** This occurs when the data is sent from the source only once and the network transmits it to multiple destinations.
- **Single-copy:** This occurs when the routing scheme promotes a single-copy forwarding strategy.
- **Multiple copies:** This occurs when the routing scheme promotes a multiple-copies forwarding strategy.
- **Compatibility with DTN:** This checks whether the underlying routing complies with any existing DTN routing scheme.
- **Routing protocols:** This specifies which underlying routing protocols can be used by the incentive scheme.
- **Communication overhead:** This specifies the method used (if any) for reducing the communication overhead.
- **Sniffing:** Ability to listen to transmissions that are destined to other entities. We use "Yes" to indicate that this feature applies to the studied incentive scheme or "No" otherwise.
- **Session key establishment:** Specifies which method is used (if any) for session key establishment.

3.4   Evaluation of Credit-Based Incentive Schemes

In this section, the following incentive schemes in mobile wireless ad hoc and challenged networks are evaluated with respect to the characteristics described in Section 3.3.

- Secure Incentive Protocol (SeIP) [25]
- Secure Multilayer Credit-based Incentive Scheme (SMART) [32]
- Express [34]
- Incentive-Compatible Opportunistic Routing (ICOR) [41]
- Incentive-Aware Routing (IAR) [42]
- Coupons [44]
- Fair Incentive Protocol (FIP) [45]
- E2-SCAN [49]

We use the symbols described in Table 3.2 in our evaluation. Our findings are captured in Tables 3.3 through 3.11. Our results are justified in the discussion that follows.

3.4.1 Secure Incentive Protocol (SeIP)

SeIP [25] is an incentive protocol implemented in a secure module residing at each node. It focuses on assigning a non-forged stamp on each packet forwarded as the proof of forwarding. Based on this, intermediate nodes are remunerated, while

Table 3.2 Symbols Used When Evaluating Incentive Schemes

Symbol	Meaning
IncCor	Incentive-compatible opportunistic routing.
IncProt	Incentive protocol.
-	The applied property is restrictive.
--	The applied property is too restrictive.
+	The applied property is desirable to a certain degree.
++	The property applies at a highest degree.
Flexible	The property can take any of the values -, +, --, ++.
O	The property is not applicable or not concerned.
Yes/No	Yes or No answer.
Sym	Symmetric role.
Asym	Asymmetric role.
Cons	Consumer.
Prov	Provider.
A	Application layer of the cooperation layered domains (Figure 3.1).
D	Discovery layer of the cooperation layered domains (Figure 3.1).
T	Transport layer of the cooperation layered domains (Figure 3.1).
N	Network layer of the cooperation layered domains (Figure 3.1).
L	Data link layer of the cooperation layered domains (Figure 3.1).
N/T	Both network and transport layers of the cooperation layered domains apply.
A/N	Both application and network layers of the cooperation layered domains apply.

(Continued)

Table 3.2 Symbols Used When Evaluating Incentive Schemes (Continued)

Symbol	Meaning
Same charge	All nodes have equal charging rate.
Proportional	Charging method where each intermediate node is awarded a certain number of credits proportional to the service it has provided.
Diff charge	Each node has a different charging rate. For example, a payment corresponding to its incurred cost, a payment proportional to its incurred cost, etc.
Application-dependent	The charging model is application dependent.
Sim	Simulation is used as the validation method.
Anal	Formal analysis/analytical models are used as the validation method.
Digital sign	Digital signature.
Public key	Public key certificate or public key cryptography.
Chain-cert	Chain of public key certificates.
Source-controlled	Source-controlled session-based approach.
Store-carry-and-forward	Store-carry-and-forward approach.
Layered conc	Layered concatenation technique.
Coop domain	Cooperative domain.
Static	Static trust.
Dynamic	Dynamic trust.

sources and destination nodes are charged with appropriate credits. Typically, a source-controlled session-based approach is employed where the source initiates an end-to-end session by unicasting a session request, which contains both the information on the rewarding frequency and a rewarding promise made by the source's secure module. Thus, SeIP stimulates cooperation at the network layer. Transport layer multicast has not yet been implemented in SeIP, but it is envisaged that SeIP can handle it.

SeIP uses a bond-based incentive pattern, where the bond is a kind of check. Therefore, the debit card or credit card principles are implemented. Thus, the remuneration type is checks. In SeIP, when an intermediate node agrees to serve on a

Table 3.3 Evaluation of Incentive Schemes with Respect to Cooperation and Transaction

Properties/ Approach	Incentive Scheme Type	Cooperation	Transaction	
		Coop. domain	Negotiation	Processing
SeIP [25]	IncProt	N	Yes	Action/ Remuneration
SMART [32]	IncProt	N/T	Yes	Action/ Remuneration
Express [34]	IncProt	N/T	O	Action/ Remuneration
ICOR [41]	IncCOR	N/T	O	Action/ Remuneration
IAR [42]	IncCOR	N	O	Action/ Remuneration
Coupons [44]	IncProt	A/N	O	Action/ Remuneration
FIP [45]	IncProt	N/T	No	Action/ Remuneration
E2-SCAN [49]	IncProt	N	O	Action/ Remuneration

session issued by the source—i.e., has agreed to provide a service on behalf of the source—it is compensated for this service by receiving from the source's tamper-proof secure module a reward at a time decided by this module, according to its promise made during the session initialization. Immediate remuneration is infeasible. Thus, the incentive pattern used by SeIP advocates a symmetric role. In addition, SeIP adopts the same charging rate for all nodes.

In SeIP, a node that participates in serving a session should receive some compensation in the form of a reward of the total credits from the source node via the source's secure module. However, a node is free to decide about its participation and can do so by not propagating the session request without bearing any punishment. Consequently, the number of entities that apply the incentive pattern (a bond-based incentive pattern in this case) may not be high. Thus, scalability is flexible—that is, it can be low (in a case where a few nodes have participated), medium, or high (in a case where a large number of nodes have participated).

The above source secure module acts as a bearer, responsible for assessing the amount of remuneration to be assigned to intermediate nodes that successfully

Table 3.4   Evaluation of Incentive Schemes with Respect to Incentive Pattern Characteristics

Properties/ Approach	Incentive Pattern Type	Role	Scalability	Remuneration Type	Remuneration Storage Site
SeIP [25]	Bond (banking)	Sym	Flexible	Checks	Provider/ Bearer
SMART [32]	Bond (bearer notes)	Sym	++	Note	Provider/ Bearer
Express [34]	Banking	Asym	+	Checks	Provider/ Bearer
ICOR [41]	Bond (banking)	Sym	++	Checks/ Virtual money/ Transfer of credits	Provider/ Bearer
IAR [42]	O	Sym	O	O	O
Coupons [44]	O	O	++	O	O
FIP [45]	Banking	Asym	++	Virtual currency	Provider/ Bearer
E2-SCAN [49]	O	O	O	O	O

fulfill the packet forwarding task. Thus, the provider/bearer is chosen as remuneration storage site.

In SeIP, for securing the charging and rewarding processes, a pairing-based method [26] involving an identity-based cryptography (IBC) is used, where the public keys of entities are directly derived from their known identifiers, thus eliminating the use of PKI-based certificates. Therefore, static trust is required, and sniffing does not apply. In addition, the tamper-proof secure module at each node is considered as a trusted entity by all nodes and can be accessed during the reward process. This secure module is also treated as a kind of credit clearance service; thus, SeIP advertises convertibility but not transferability.

Finally, SeIP addresses selfishness only and can deal with attacks 1, 2, 3, and 4. Also, SeIP assumes the discovery of a predefined path between source and destination nodes (for instance, by using a dedicated routing protocol). Thus, the routing mechanism underlying the SeIP protocol is not compatible with DTN routing.

Table 3.5   Evaluation of Incentive Schemes with Respect to Security and Trust Models

Properties/ Approach	Trust Usefulness	Trust Type	Trustee	Anonymity	Content Integrity Check
SeIP [25]	No	Static	TPSM [25]	No	TPSM [25]
SMART [32]	No	Static	Cons.	Yes	Digital sign
Express [34]	No	Static	Cons./Bank	O	Hash chains
ICOR [41]	No	Static	Cons./Bank	O	O
IAR [42]	No	O	O	O	Digital sign
Coupons [44]	O	O	O	O	O
FIP [45]	No	Static	Cons./Bank	Yes	Digital sign [46,47]
E2-SCAN [49]	O	Static	O	O	End-to-end and link encryption

3.4.2 The Secure Multilayer Credit-Based Incentive Scheme (SMART)

The SMART scheme [32] incorporates both the single-copy and the multicopy data forwarding strategies to ensure the transfer of bundle messages. Thus, SMART supports both unicast and transport layer multicast. In addition, SMART introduces a layer concatenation technique, where a layered coin provides virtual electronic credits as an incentive to stimulate cooperation among nodes while proving its effectiveness by restraining selfish behavior on the network layer. The latter is achieved through its rewarding and charging mechanisms implemented via a profit-sharing model.

SMART uses a bond-based incentive pattern, where the bond is a bearer note. Indeed, in SMART, a forwarding node (or provider) uses a digital signature as an agreement that it will provide a service—i.e., will fulfill the forwarding of the received packets—to its predecessor (our so-called consumer) based on a predefined class of service requirement. In return, it will be rewarded implicitly by the consumer through a TTP entity (here, the virtual bank [VB] that acts as a bearer) according to the reward policy in the future. Thus, the symmetric role applies.

In SMART, a node that correctly fulfills its packet forwarding task should receive some compensation in the form of a dividend of the total credits from the source node via the VB. Hence, nodes are naturally motivated to participate in

Table 3.6   Evaluation of Incentive Schemes with Respect to Implementation and Performance

Properties/ Approach	Implementation		Performance		
	With Respect to Cryptography	With Respect to Tamper Resistance	With Respect to Validation Method	With Respect to Mobility Model	With Respect to Network Models
SeIP [25]	IBC	TPSM [25]	Sim	Modified random waypoint	General MANETs
SMART [32]	Public key	No	Sim	Map-based mobility model	Vehicular DTNs, OCUNs
Express [34]	Public key	No	Anal	Notc	MANETs
ICOR [41]	Public key	No	Sim	Notc	OCUNs
IAR [42]	Public key	No	Sim	Mobility traces	DTNs, OCUNs
Coupons [44]	Public key	O	Sim	Random waypoint, Manhattan Grid, Cluster-based mobility models	DTNs, Disconnected ad hoc networks, OCUNs
FIP [45]	Public key	No	Anal	Notc	MANETs
E2-SCAN [49]	Public key	O	Sim	Modified random waypoint	MANETs

packet forwarding in order to gain as many credits as possible. Consequently, the number of entities that apply the incentive pattern is high. This has been validated through the results obtained on the stability of the network throughput [32]. Hence, SMART can successfully stimulate a large number of selfish nodes to participate in packet forwarding. Thus, scalability applies at the highest degree.

SMART relies on accessibility of the VB to run credit clearance. The amount of remuneration provided to a node is assessed by this VB based on collected layered

Table 3.7   Evaluation of Incentive Schemes with Respect to Payment and Reward Models

Properties/ Approach	Transferability	Convertibility	Charging Model	Charging Rate
SeIP [25]	No	Yes	Proportional	Same charge
SMART [32]	No	No	Profit sharing	Diff. charge
Express [34]	No	Yes	Game theoretical-based	Diff. charge
ICOR [41]	No	Yes	Game theoretical-based	Application-dependent
IAR [42]	O	O	Game theoretical-based TFT	Same charge
Coupons [44]	O	O	Pyramid-like or Flat	Application-dependent
FIP [45]	No	Yes	Not specified	Diff. charge
E2-SCAN [49]	O	O	Secret sharing techniques [51]	Diff. charge

coins received from that node. The VB then uses a profit-sharing model to determine the amount of credits to be rewarded to a node; thus, SMART allows flexible remuneration.

In SMART, security is enforced by the use of hash functions in the design of PKI-based certificates. Each node has a unique public key certificate assigned by an offline security manager (OSM). Thus a chain of certificates (i.e., signatures) is constructed when designing the layered chain, which in turn determines the secure path to be used for packet forwarding. Thus, trust is static. In addition, trust is implemented via a trusted third party (here the VB), and the participating entities do not have to disclose their identities. For these reasons, SMART supports anonymity as well as a public key–based cryptography infrastructure.

SMART deploys a digital signature scheme as the trust mechanism available to nodes. This is then used to track down the propagation path, thereby ensuring the security of the data forwarding. Hence, content integrity is established through non-forgeable digital signatures. Finally, SMART has been designed to protect against attacks 1, 7, 8, and 9.

Table 3.8   Evaluation of Incentive Schemes with Respect to Payment and Reward Models

Properties/ Approach	Tamper-Proofness	Flexible Remuneration	Aggregation of Remuneration
SeIP [25]	Via the TPSM [25]	Notc	Yes
SMART [32]	Chain of cert.	Yes	Yes
Express [34]	Cert.	Yes	No
ICOR [41]	Secret key cryptography [40]	Yes	Yes
IAR [42]	IBC [43]	O	O
Coupons [44]	Notc	O	O
FIP [45]	Via theTCCS [45]	Yes	No
E2-SCAN [49]	Notc	O	O

3.4.3 The Express Protocol

The Express protocol [34] can be viewed as an enhancement of SPRITE [35], where the incentive pattern (here the banking pattern) is geared toward route discovery. In Express, security is enforced by the use of hash chains (the product of successive applications of a hash function) to protect against fraudulent selfish nodes on the

Table 3.9   Evaluation of Incentive Schemes with Respect to Node Misbehaviors and Attack Models

Properties/Approach	Node Misbehavior	Attack Models
SeIP [25]	Selfishness	Attacks 1, 2, 3, 4
SMART [32]	Selfishness, mass selfishness, individual selfishness	Attacks 1, 7, 8, 9
Express [34]	Selfishness, lavishness	Attacks 5, 10, 11, 12
ICOR [41]	Selfishness, lavishness	Attacks 9, 10, 11, 12
IAR [42]	Selfishness	Attack 13
Coupons [44]	Maliciousness	Attacks 14, 15
FIP [45]	Selfishness	Attacks 2, 11
E2-SCAN [49]	Maliciousness	Attacks 16, 17

Table 3.10 Evaluation of Incentive Schemes with Respect to Data Dissemination Models

Properties/ Approach	Packet Forwarding Strategy	Unicast	Multicast	Single-copy	Multiple-copy
SeIP [25]	Source controlled	Yes	No	Yes	No
SMART [32]	Layered conc. [32]	Yes	Yes	Yes	Yes
Express [34]	Source controlled	No	Yes	Yes	No
ICOR [41]	Store-carry-and-forward	Yes	Yes	Yes	O
IAR [42]	Pairwise TFT	O	O	O	O
Coupons [44]	Store-carry-and-forward	O	O	O	O
FIP [45]	Source controlled	Yes	Yes	Yes	O
E2-SCAN [49]	Source controlled	No	Yes	Yes	O

network layer while supporting transport layer multicast. Thus, Express stimulates cooperation at both the network and transport layers. In addition, hash chains are also used to reduce the computational overhead of nodes.

Express relies on the accessibility of a dedicated banker node (here the reliable clearance center, or RCC), which uses a single-copy forwarding technique to assign credits and remuneration to nodes based on reports it receives from the network on nodes' activities. Based on these reports, the RCC (which handles the roles of credit manager and digital certificate issuer) judges the cooperativeness of nodes and secures the micropayment by assigning the appropriate amount of remuneration to each node. Therefore, negotiation is dispensable. It is also assumed that all nodes are able to access the RCC. In addition, the RCC is trusted in terms of credit balance management; thus, transferability of checks is not considered.

In Express, the source node uses the aforementioned public key-based digital certificates to identify intermediate nodes to which packets should be forwarded; that is, to identify only those nodes that hold valid certificates from the RCC, and to track down the forwarding path.

Table 3.11 Evaluation of Incentive Schemes: Data Dissemination Models

Properties/ Approach	Compatibility with DTN	Routing Protocol	Communication Overhead	Sniffing	Session Key Establishment
SeIP [25	No	Ariadne [28], SecMR [30]	Bloom filter method [27]	O	Pairing-based method
SMART [32]	Yes	DTN-based, Epidemic-like, Probabilistic spray and wait	Digital sign. as aggregation hash chains signature	O	Pairing-based method
Express [34]	No	Power-aware-based (e.g. MTPR [36], MMBCR [37], CMMBCR [38]	Hash chains	O	Hash chains
ICOR [41]	No	MORE [39], DSR [18]	O	Yes	Secret-key cryptography [40]
IAR [42]	Yes	OSPF [42]	O	O	HIBC [43]
Coupons [44]	Yes	Epidemic-like [48]	O	O	O
FIP [45]	No	AODV [45]	O	O	O
E2-SCAN [49]	No	Modified AODV	Secret sharing techniques [51]	Yes	O

Express uses hash chains on messages to defend against cheating by nodes. To accomplish this, a charging/payment model, in the form of a game theoretical-based analytical model, is used to bring a higher payoff to selfish nodes, discouraging them from benefiting from cheating. In other words, Express provides appropriate incentives and fines so that rational nodes do not prefer to misbehave, thus encouraging a large number of nodes to participate in cooperation enforcement. Hence, scalability applies, but at a certain degree because no experimental or simulation performance results have been provided in [34] to sustain the analytical model and to assess the scalability rate. Finally, Express can protect against attacks 5, 10, 11, and 12.

3.4.4 Incentive-Compatible Opportunistic Routing (ICOR)

The ICOR scheme proposed by Wu et al. [41] is the first opportunistic routing-based protocol that advocates the use of incentives, in the form of credits, as a stimulus for packet forwarding by encouraging each user node to honestly participate in network routing despite opportunistic transmissions. Thus, the ICOR protocol stimulates cooperation at the network layer. This protocol also takes advantage of broadcast transmissions to send information through multiple concurrent relays, and thus transport layer multicast is also supported.

In ICOR, when a source node initiates a session with a destination node, any intermediate node involved in the propagation path is expected to receive a payment (real money or a transfer of credit) in the form of a reward, granted by the source node in recognition of its service in participating in the bundle forwarding. This payment is not immediate, but rather is accumulated for the duration of the session, and the node is expected to receive the total payment for the entire session. Thus, ICOR relies on a bond-based incentive pattern, where the bond is a banking pattern. Aggregation of remuneration does apply, since the payment of a well-behaved node is not immediate.

ICOR relies on the accessibility of a banker node (a so-called *routing decision maker*, or RDM) that acts as a credit clearance entity by collecting the information of loss probabilities at each node and by using this information to compute the amount of payment (i.e., reward) to be awarded to intermediate nodes by the source node. This payment breaks into two components: data transmissions reward and auxiliary transmissions reward. The latter constitutes the incentive for routing, in the sense that well-behaved nodes (i.e., nodes that have faithfully reported the loss probabilities of their outgoing links) are assigned auxiliary transmission rewards in addition to their already granted data transmission rewards. To this effect, a technique referred to as *strict dominant strategy equilibrium* is used to motivate each node to honestly report its loss probabilities. Nodes that correctly fulfill their packet forwarding tasks should receive a reward from the source node via the RDM (which has already assessed the amount of remuneration). This is accomplished through a single control message transmitted in a sprite-like manner so that the negotiation phase becomes dispensable. Thus, the symmetric role applies and ICOR promotes action and remuneration.

In ICOR, the reward mechanism is application-dependent and can be in the form of virtual money having a cash value or transfer of credits; for instance, via checks. Due to the role played by the RDM (it acts as a bearer), transferability and tamper resistance do not apply. However, convertibility is ensured.

In ICOR, nodes are motivated to participate in packet forwarding to earn as many auxiliary transmission rewards as possible. This incentive is implemented through a game theoretical-based technique (so-called *strict dominant strategy equilibrium*), which imposes a severe punishment on detected cheating behaviors—nodes that do not honestly report their loss probabilities are highly punished. Its effectiveness is rigorously demonstrated analytically and assessed experimentally. Thus, ICOR promotes scalability at a high degree while restraining selfish behaviors.

In ICOR, cheating refers to the fact that a selfish node can deviate from the protocol. This behavior is prevented by using the above strict dominant strategy equilibrium to encourage nodes to report their loss rates, coupled with a method used by the source node to compute these loss rates. Lavishness occurs when a node deliberately fails to report its loss rates while taking advantage of the routing process to acquire some credits fraudulently. Thus, selfishness and lavishness both apply.

In ICOR, routing decisions and their realizations are secured using PKI-based cryptographic techniques such as CORSAC [40] to address the combined problem of routing and packet forwarding. In addition node overhearing occurs, since the protocol enhances opportunistic routing, and thus sniffing applies. Finally, ICOR can protect against attacks 9, 10, 11, and 12.

3.4.5 Incentive-Aware Routing (IAR)

The incentive-aware routing (IAR) mechanism proposed by Shevade et al. [42] is the first practical TFT-based incentive mechanism for DTNs. In this scheme (referred to as TFT-for-DTN), TFT uses a DTN routing to optimize the routes when all nodes in a DTN are cooperative, as well as a selfish DTN routing to allow selfish nodes to optimize their own individual performances while conforming to TFT constraints. In the latter, every node uses a kind of Open Shortest Path First (OSPF)-based link state routing to keep track of the information about links in the network. Thus, TFT-for-DTN is designed to restrain selfish behavior on the network layer.

In TFT-for-DTN, TFT reciprocates good or bad behavior only between neighbors. Packet acknowledgment is a sign used by a node to justify that the next hop has successfully achieved its packet forwarding task. This feedback allows a node to engage in balanced exchange with its neighbors, and therefore to reward good behavior with equal reciprocal service while ignoring misbehavior. Thus, TFT-for-DTN advocates a symmetric role.

As a TFT-based incentive mechanism, TFT-for-DTN does not require trusted nodes or special hardware. However, in TFT-for-DTN, when using the selfish DTN routing to send packets from source to destination, the source route is digitally signed by the sender—for instance, via a hierarchical identity-based cryptography (HIBC). Thus, digital signatures are used for content integrity checks and for ensuring the security of the forwarding path.

3.4.6 Coupons

The Coupons scheme [44] is quite different from conventional incentive schemes, as it does not address the issue of incentives in its general form as most schemes do, but from an application scenario perspective. Here, the incentive is contextual and is based on an ordered list of unique IDs appended to messages. Thus, Coupons stimulates cooperation on the application layer.

The Coupons scheme also focuses on data sharing through opportunistic contact. It is designed for providing content sharing by stimulating adaptive localized interactions between potential users. Data dissemination is achieved by implementing some mechanism on top of basic flooding. Typically, the information is continuously broadcast as a user (assumed to be independent) moves and comes into contact with another user. A positive acknowledgment from this user triggers the transmission of the coupon via a three-way handshake-based pull model or preconfigured user profiles. Users are stimulated in relaying their content in a pyramid-like manner, up to the destination. A feedback-based back-off mechanism is implemented to achieve bundle forwarding. Thus, Coupons promote cooperation at the network layer too.

In Coupons, every node (i.e., independent user) is assigned a unique ID, then shares a coupon as it comes in contact with an immediate surrounding neighbor, building an ordered list of unique IDs appended to a message, which in turn determines the forwarding path to the destination. A node that successfully participates in the packet forwarding is rewarded a credit (in the form of a coupon). The users residing at the top of the pyramid receive more credits than those residing at the bottom. Nodes are naturally motivated to participate in packet forwarding to earn as many credits as possible. Thus, scalability applies at a highest degree. Coupons can efficiently deal with attacks 14 and 15.

3.4.7 Fair Incentive Protocol (FIP)

FIP [45] lays stress on network-level selfishness. The underlying protocol routing (AODV) also supports transport layer multicast. Thus, FIP promotes cooperation at both the network and transport layers, by providing incentives for nodes to faithfully forward packets. FIP also assumes the existence of a trusted third party (a so-called Trusted Credit Clearance Service, or TCCS) to run the credit clearance service, where credits are considered as virtual currency (which could be real money). Thus, the banking pattern and convertibility both apply.

In FIP, each node (with a credit account) that successfully participates in the packet forwarding of other nodes will receive some compensation from the TCCS (in the form of credits), but with no explicit rule imposed on the beneficiary node, and thus an asymmetric role applies.

In FIP, the source sends a packet along with an incentive credit message. Each intermediate node relays the packet as well as the credit message, along with some kind of public key-based signature to prove that it is participating in the packet forwarding. At the end of the forwarding process, the destination node computes a receipt (signature), which is reported to the TCCS. The TCCS uses it to assess the amount of credits (from the source node's account) that should be allocated to each intermediate node and the last intermediate node. In order to be allocated a credit value, an intermediate node should have received an authorization, in the form of a receipt, from the destination node. This form of stimulus has been validated by simulation [45], showing that scalability is highly enforced in FIP.

In FIP, since the TCCS determines the amount of remuneration to be assigned to a node, transaction negotiation is not required. In addition, the security of message is enforced by cryptographic operations involving some secure short signature schemes [46,47]. Thus, FIP does not require any tamper-resistant device. Finally, FIP can protect against attacks 2 and 11.

3.4.8 E2-SCAN Protocol

The E2-SCAN incentive protocol [49] is a network layered security solution designed to monitor both routing and packet forwarding activities at each node. For the node's monitoring purpose, it uses a packet drop–detection algorithm mechanism similar to the watchdog technique [50], but with a distinct collaborative monitoring mechanism. Therefore, each node monitors the packet forwarding activity of its neighbors by overhearing the channel and comparing ongoing data transmission with previously recorded routing messages. Thus, sniffing applies.

In E2-SCAN, when a new node joins the network, it is assigned a token with a short lifetime. The aim of the node is to keep a longer token lifetime. A charging model [51] is used to evaluate the token lifetime that decreases the token renewal overhead with time. Depending on the node's behavior, its credit is increased slowly (in case of legitimate nodes) or decreased (in case of malicious nodes). This helps in quickly detecting the malicious nodes by isolating them from the network. More credits accumulated means longer token lifetime, which in turn means less frequent renewal of the lifetime token.

E2-SCAN does not assume that nodes are equipped with an a priori trust relationship. Instead, it exploits some polynomial secret sharing techniques [51] as an alternative for a trust mechanism to enhance the tolerance and collaboration among nodes. E2-SCAN does not apply any cryptographic primitives on routing messages. Instead, asymmetric cryptography, such as public-key cryptographic primitives, is used to prevent forgery of tokens. E2-SCAN addresses only the maliciousness of a node by protecting against attacks 16 and 17.

3.5 Conclusions and Future Trends

In this chapter, we have identified some of the most recently proposed credit-based incentive schemes for stimulating cooperation in mobile wireless ad hoc networks and challenged networks. We examined these schemes based on a set of well-known characteristics of incentive schemes, highlighting those that are tailored to opportunistic networks. Extensive qualitative comparisons of these schemes have been conducted using various parameters and attributes. It would be interesting to pursue this study by investigating other types of incentive schemes. Existing trust- and reputation-based solutions proposed for mobile networks and challenged networks can be extended for use in mobile and wireless opportunistic networking services and applications.

References

[1] P. Obreiter and J. Nimis. A taxonomy of incentive patterns: The design space of incentives for cooperation. *Lecture Notes in Computer Science,* 2872:89–100, 2005.

[2] P. Obreiter, B. Konig-Ries, and M. Klein. Stimulating cooperative behavior of autonomous devices: An analysis of requirements and existing approaches, In *Proceedings of 2nd International Workshop on Wireless Information Systems (WIS 2003),* 2003.

[3] L. Lilien, Z. H. Kamal, A. Gupta, I. Woungang, and E. Bonilla Tamez. Quality of service in an opportunistic capability utilization network, In *Mobile Opportunistic Networks: Architectures, Protocols, and Applications,* M. K. Denko (ed.), Auerbach Publications, Taylor & Francis Group, Boca Raton, Florida.

[4] Haggle, A. European Union funded project in situated and autonomic communications, http://www.haggleproject.org/index.php/Main_Page, accessed October 23, 2009.

[5] The iClouds Project. Opportunistic communication among people, http://iclouds. tk.informatik.tu-darmstadt.de/, accessed October 23, 2009.

[6] N. B. Chang and M. Liu. Competitive analysis of opportunistic spectrum access strategies, In *Proceedings of IEEE 27th Conference on Computer Communications at INFOCOM 08,* pp. 1535–1542, Phoenix, Arizona, April 2008.

[7] C. Boldrini, M. Conti, and A. Passarella. ContentPlace: Social-aware data dissemination in opportunistic networks. In *Proceedings of the 11th Intern. Symposium on Modeling, Analysis, and Simulation of Wireless and Mobile Systems,* pp. 203–210, Vancouver, Canada, 2008.

[8] C. Boldrini, M. Conti, and A. Passarella. Modeling data dissemination in opportunistic networks. In *Proceedings of Challenged Networks 2008,* pp. 89–96, 2008.

[9] L. Pelusi, A. Passarella, and M. Conti. Opportunistic networking: Data forwarding in disconnected mobile ad hoc networks. *IEEE Communications Magazine,* 44(11): 2006, pp. 134–141.

[10] L-J. Chen, C-H. Yu, C-L. Tseng, H-H. Chu, and C-F. Chou. A content-centric framework for effective data dissemination in opportunistic networks. *IEEE Journal of Selected Areas in Communications (JSAC),* 26(5):761–772, June 2008.

[11] O.V. Drugan, T. Plagemann, E. and Munthe-Kaas. Building resource aware middleware services over MANET for rescue and emergency applications. In *Proceedings of the 16th Annual IEEE Intl. Symposium on Personal Indoor and Mobile Radio Communications,* Berlin, Germany, September 2005.

[12] L. Lilien, Z. H. Kamal, and A. Gupta. Opportunistic networks: Research challenges in specializing the P2P paradigm. In *Proceedngs of the 3rd International Workshop on P2P Data Management, Security, and Trust (PDMST'06),* pp. 722–726, Krakow, Poland, September 2006.

[13] Newcom++. State of the art of research on opportunistic networks, and definition of a common framework for reference models and performance metrics. *Technical Report, NewCom,* http://www.newcom-project.eu:8080/Plone/public-deliverables/DR11.1-final-1.pdf/view, accessed October 23, 2009.

[14] P. Hui, A. Chaintreau, J. Gass, J. Scott, J. Crowcroft, and C. Diot. Pocket switched networking: Challenges, feasibility, and implementation issues. In *Proceedings of WAC 05,* Athens, Greece, October 2005.

[15] ANA Project. Autonomic network architecture, http://www.ana-project.org, accessed on October 23, 2009.

[16] E. Yoneki, P. Hui, S. Chan, and J. Crowcroft. A socio-aware overlay for publish/subscribe communication in delay tolerant networks. In *Proceedings of MSWiM 2007,* Crete, Greece, October 2007.

[17] C. Boldrini, M. Conti, I. Iacopini, and A. Passarella. Exploiting users' social relations to forward data in opportunistic networks: The HiBOp solution. *Pervasive and Mobile Computing* 4:633–657, 2008.

[18] D. B. Johnson and D. A. Malt. Dynamic source routing in ad hoc wireless networks. In *Mobile Computing,* Tomasz Imielinski and Hank Korth (eds.), 1996. Accessed October 23, 2009, http://www.monarch.cs.cmu.edu/monarch-papers/kluwer-adhoc.ps.

[19] C. Caini, P. Cornice, R. Firrincieli, and D. Lacamera. A DTN approach to satellite communications. *IEEE Journal of Selected Areas in Communications* 26(5): June 2008, pp. 820–827.

[20] Z. Zhang. Routing in intermittently connected mobile ad hoc networks and delay tolerant networks: Overview and challenges. *IEEE Communication Surveys and Tutorials,* 8(1) March 2007, pp. 24–37.

[21] L. Lilien. A taxonomy of specialized ad hoc networks and systems for emergency applications. In *Proceedings of the 1st International Workshop on Mobile and Ubiquitous Context Aware Systems and Applications (MUBICA 2007),* Philadelphia, Pennsylvania, August 2007.

[22] V. Cerf, S. Burleigh, A. Hooke, L. Torgerson, R. Durst, K. Scott, K. Fall, and H. Weiss. Delay tolerant networking architecture, July 2004, Internet draft, http://www.rfc-editor.org/rfc/rfc4838.txt, accessed October 23, 2009.

[23] H. Janzadeh, K. Fayazbakhsh, B. Bakhshi, and M. Dehghan. A novel incentive-based and hardware-independent cooperation mechanism for MANETs. In *Proceedings of IEEE Wireless Communications and Networking Conf. (WCNC 2008),* pp. 2462–2467, Las Vegas, Nevada, March–April 2008.

[24] M. Zghaibeh and F. C. Harmantzis. Lottery-based pricing scheme for peer-to-peer networks. In *Proceedings of the IEEE International Conference on Communications,* Vol. 2, pp. 903–908, June 2006.

[25] Y. Zhang, W. Lou, W. Liu, and Y. Fang. A secure incentive protocol for mobile ad hoc networks. *ACM Wireless Networks,* 13(5):569–582, October 2007.

[26] D. Boneh and M. Franklin. Identity-based encryption from the weil pairing. In *Proceedings of CRYPTO'01, LNCS,* Vol. 2139, pp. 213–229, Springer-Verlag, 2001.

[27] B. Bloom. Space/time trade-offs in hash coding with allowable errors. *Communications of the ACM,* 13(7): July 1970, pp. 422–426.

[28] Y. C. Hu, A. Perrig, and D. B. Johnson. Ariadne: A secure on-demand routing protocol for ad hoc networks, *ACM MOBICOM,* Atlanta, Georgia, September 2002.

[29] K. Sanzgiri, D. LaFlamme, B. Dahill, B. Levine, C. Shields, and E. B. Royer. Authenticated routing for ad hoc networks, *IEEE Journal of Selected Areas Communication,* 23(3):598–610, March 2005.

[30] P. Kotzanikolaou, R. Mavropodi, and C. Douligeris. Secure multipath routing for mobile ad hoc networks. In *Proceedings of the 2nd Annual Conference on Wireless On-Demand Network Systems and Services (WONS'05)*, St. Moritz, Switzerland, January 2005.

[31] D. Boneh and M. Franklin. Efficient generation of shared RSA keys, *Journal of the ACM (JACM)*, 48(4):702–722, July 2001.

[32] H. Zhu, X. Lin, R. Lu, Y. Fan, and X. Shen. SMART: A secure multi-layer credit based incentive scheme for delay-tolerant networks, *IEEE Trans. on Vehicular Technology* Vol. 58, no. 8, 2009, pp. 4628–4639.

[33] D. Boneh and M. Franklin. Boneh-Franklin identity based encryption revisited. In *Lecture Notes in Computer Science*, 3580:791–802, 2005.

[34] H. Janzadeh, K. Fayazbakhsh, M. Dehghan, and M. S. Fallah. A secure credit-based cooperation stimulating mechanism for MANETs using hash chains. *Future Generation Computer Systems*, 25(8):926–934, 2009.

[35] S. Zhong, J. Chen, and Y. R. Yang. Sprite: A simple, cheat-proof credit-based system for mobile ad hoc networks. In *Proceedings of IEEE INFOCOM 2003*, pp. 1987–1997, San Francisco, California, 2003.

[36] K. Scott and N. Bambos. Routing and channel assignment for low power transmission in PCS. In *Proceedings of IEEE ICUPC*, 1996.

[37] S. Singh, M. Woo, and C. S. Raghavendra. Power-aware routing in mobile ad hoc networks. In *Proceedings of the 4th Annual ACM/IEEE International Conference on Mobile Computing and Networking (ACM MoBiCom)*, Dallas, TX, USA, pp. 181–190, 1998.

[38] C. K. Toh. Maximum battery life routing to support ubiquitous mobile computing in wireless ad hoc networks. *IEEE Communications Magazine*, 39(6):138–147, 2001.

[39] S. Chachulski, M. Jennings, S. Katti, and D. Katabi. Trading structure for randomness in wireless opportunistic routing. In *Proceedings of ACM SIGCOMM'07*, Kyoto, Japan, August 2007.

[40] S. Zhong, L. Li, Y. Liu, and Y. R. Yang. On designing incentive-compatible routing and forwarding protocols in wireless ad-hoc networks: An integrated approach using game theoretical and cryptographic techniques. In *Proceedings of the 11th International Conference on Mobile Computing and Networking (MobiCom'05)*, Cologne, Germany, September 2005.

[41] Fan Wu, Tingting Chen, Sheng Zhong, L. Erran Li, and Richard Yang. Incentive-compatible opportunistic routing for wireless networks. In *Proceedings of ACM MOBICOM*, San Francisco, California, September 2008.

[42] U. Shevade, H. H. Song, L. Qiu, and Y. Zhang. Incentive-aware routing in DTNs. In *Proceedings of the 16th EEE International Conference on Network Protocols (ICNP 2008)*, pp. 238–247, Orlando, Florida, October 19–22, 2008.

[43] A. Seth and S. Keshav. Practical security for disconnected nodes, In *Proceedings of the 1st ICNP Workshop on Secure Network Protocols (NPSec 2005)*, Boston, MA, USA, pp. 31–36, 2005.

[44] A. Garyfalos and K. C. Almeroth. Coupons: A multilevel incentive scheme for information dissemination in mobile networks. *IEEE Transactions on Mobile Computing*, 7(6):June 2008, pp. 792–804.

[45] R. Lu, X. Lin, H. Zhu, C. Zhang, P.H. Ho, and X. Shen. A novel fair incentive protocol for mobile ad hoc networks. In *Proceedings of IEEE WCNC'08*, Las Vegas, Nevada March 31–April 3, 2008.

[46] F. Zhang, R. Safavi-Nani, and W. Susilo. An efficient signature scheme from bilinear pairings and its applications. In *Proceedings PKC 2004, LNCS 2947*, pp. 277–290, Springer-Verlag, 2004.

[47] J. Camenisch, S. Hohenberger, and M. Pedersen. Batch verification of short signatures. In *Advances in Cryptology—EUROCRYPT 2007, LNCS Vol. 4515*, pp. 246–263, 2007.

[48] A. Vahdat and D. Becker. Epidemic routing for partially connected ad hoc networks. In *Technical Report CS-200006*, Duke University, April 2000.

[49] S. K. Dhurandher, S. Misra, S. Ahlawat, N. Gupta, and N. Gupta. E2-SCAN: An extended credit strategy-based energy-efficient security scheme for wireless ad hoc networks. *IET Communications Journal* (U.K.), 3(5):809–819, 2009.

[50] S. Marti, T. Giuli, K. Lai, and M. Baker. Mitigating routing misbehavior in mobile ad hoc networks. In *Proceedings of ACM MOBICOM*, Boston, USA, pp. 255–265, 2000.

[51] J. Kong, P. Zerfos, H. Luo, S. Lu, and L. Zhang. Providing robust and ubiquitous security support for MANET. In *Proceedings of IEEE ICNP*, pp. 251–260, 2001.

[52] Bundle protocol specifications, http://www.rfc-archive.org/getrfc.php?rfc=5050, accessed October 23, 2009.

Chapter 4

Opportunism in Mobile Ad Hoc Networking

Marcello Caleffi
University of Naples Federico II Naples, Italy

Luigi Paura
University of Naples Federico II Naples, Italy

Contents

4.1 Introduction

In the last two decades, great attention has been devoted to the *ad hoc network-ing* paradigm, and there are a large number of routing protocols designed for it. These protocols cover a wide range of design choices and approaches, from simple modifications of traditional solutions for wired networks to more innovative and complex schemes. Most of these protocols [9,12,16,27,28] are based on the *multi-hop* paradigm, which allows nodes to extend the limited coverage range of wireless communications by exploiting neighbors as cooperative relays.

In fact, direct forwarding allows nodes to communicate only if they are within the direct transmission range of each other. With reference to the simple topology shown in Figure 4.1, if nodes *s* and *d* have to communicate and the link quality is poor, they cannot communicate at all. In contrast, if the network layer adopts mul-tihop forwarding, a pair of nodes can also communicate if they are not within the direct transmission range of each other or if their link quality is poor. With refer-ence to the previous example (Figure 4.2), the neighbor *r* allows *s* to communicate with *d* acting as a relay—that is, by storing in its buffer the packets received from *s* and by sending them to *d*. Clearly, multihop communication can involve multiple relays, and in such a case the step above is repeated until the packet is received at the destination.

For more than a decade, multihop forwarding has been considered a suitable strategy for networking in ad hoc networks, since it fits well in scenarios character-ized by dynamic topology with no available infrastructure or central management. However, the main issue with multihop routing is that it tries to *fortify* the scenario so that it behaves like a wired network instead of exploiting the key features of wire-less technology: the broadcasting and the unreliability.

In fact, multihop routing completely hides the broadcast nature of wireless communications in data forwarding by imposing a requirement at the data-link layer that nodes have to discard data packets not directly sent to them, although they have correctly received such packets. Moreover, it usually counteracts the

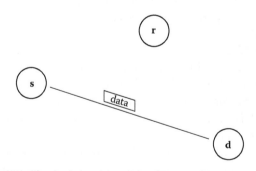

Figure 4.1 Direct forwarding.

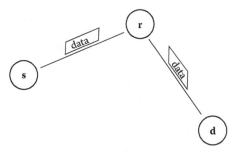

Figure 4.2 Multihop forwarding.

time-variant impairment of the wireless propagation by means of Automatic Repeat Request (ARQ) or Forward Error Control (FEC) data-link techniques.

As opposed to *fortifying* the environment, a concept recently proposed is to exploit the good nature of wireless communications, namely the broadcasting, to compensate for the unreliability. This design philosophy—opportunistic routing—aims at relaxing the assumption that the wireless propagation conditions are stationary enough so that they allow a persistent communication among neighboring nodes. In such a way, opportunistic routing is able to provide connectivity in scenarios where traditional ad hoc networking fails. This chapter describes the fundamental characteristics of opportunistic routing (Section 4.2), along with the main features of some representative routing protocols belonging to this class (Section 4.3). The routing protocols have been selected for one or more of the following reasons: (1) they are popular choices in the research community; (2) they may be interesting, illustrative examples of this class; (3) they may have unique features that make them interesting. In the following we do not make comparisons among the considered protocols, since there are many published performance comparisons [20,23,30,35–38]. Moreover, the citations for the considered protocols themselves often provide performance evaluations of the protocol.

4.2 Opportunistic Routing

As mentioned in the previous paragraph, opportunistic routing is a class of routing protocols that, rather than counteracting, tries to take advantage of the time-variant nature of the environment to provide end-to-end connectivity in scenarios where traditional networking fails. Let us provide an example of how opportunistic routing works by comparing it with traditional ad hoc routing.

If the network layer adopts the multihop routing paradigm, data communications are unicast. Therefore, the next hop for each packet has to be singled out before sending that packet on the link—i.e., the next hop selection happens at the sender side. Clearly, this strategy is not optimal since routing progress (i.e., the progress of a data packet toward the destination) is achieved only if the packet is received by the selected next hop. In other words, traditional ad hoc routing is unable to exploit the opportunity offered by an unselected relay closer to the destination than the source.

With reference to the example reported in the previous paragraph (Figure 4.2), where the source *s* has to communicate with *d* and *r* is a neighbor of *s* that is closer* to *d* than *s* itself, the source can select *r* or *d* as the next hop. If we suppose that *s* selects *r*—i.e., it selects the most reliable link—it will be unable to take advantage of favorable wireless propagation conditions that allow *d* to receive the packet. If we suppose that *d* has been selected as the next hop and that the packet reaches *r* but not *d*, again no routing progress is made. In contrast, opportunistic routing requires that the source simply broadcast the data packets without worrying about next hop selection, which happens at the receiver side. In such a way, routing progress is achieved every time that a node closer to the destination than the source receives the packet, and that node becomes the next hop. With reference to the above example, if the packet is received by the destination—i.e., if favorable propagation conditions exist—significant routing progress is achieved. However, if only node *r* receives the packet, it becomes responsible for packet forwarding and routing progress is achieved anyway. Clearly, if neither *r* nor *d* receive the packet, *s* retains the responsibility for packet forwarding.

We note that ad hoc networks exhibit an inherent distributed spatial diversity, due to both node mobility and wireless propagation instability. In fact, with respect to node mobility, an opportunity happens when the selected next hop moves outside the sender's transmission range (and the packet is received by a less favorable neighbor), or when a packet is received by a more favorable neighbor that has moved into the sender's transmission range but has not yet been recognized as a neighbor by the neighbor discovery mechanism. Moreover, in wireless propagation, an opportunity happens when the propagation conditions of the selected link worsen during or just before the packet transmission (and the packet is received by a less favorable neighbor), or when a packet is received by a more favorable neighbor along a link previously classified as unreliable.

Opportunistic routing allows nodes to exploit such opportunities to increase the probability that packets progress toward their destinations as covered in Section 4.2.1. However, these opportunities are not "free" since they require coordination among nodes, which means routing overhead as will be pointed out in Section 4.2.2.

---

* Here and in the following text, we use the term closer to indicate that a node is a preferable forwarder with respect to another node.

4.2.1 Benefits

Opportunistic routing differs from traditional ad hoc routing in that it exploits the broadcast nature of a wireless medium by deferring the route selection to the receiver side. Clearly, this feature copes well with unreliable and unpredictable wireless transmission, and in this section we describe the two main advantages of opportunistic routing in the presence of unreliable links—*multiuser diversity* and *route opportunism*.

In the following we assume the presence of perfect coordination among the nodes and we neglect the additional overhead introduced by opportunistic routing. Such issues will be discussed in the next section.

4.2.1.1 Multiuser Diversity

Opportunistic routing exploits multiuser diversity—that is, the availability of multiple neighbors whose links can be modeled as statistically independent channels [38] in order to manage the unreliability of wireless communications. As an example, consider the diamond topology depicted in Figure 4.4, where the source s can reach the destination d through five relays, r_i, and $d_{i,j}$ is the average delivery ratio of the packets sent by i to j:

$$d_{s,r_i} = 0.2 \quad \forall i \tag{4.1}$$

$$d_{r_i,d} = 1.0 \quad \forall i. \tag{4.2}$$

Since traditional ad hoc routing selects the next hop at the sender side, it is unable to take advantage of a transmission that reaches a node other than the selected one. So, the average number of link transmissions $\bar{n}_{s,d}(s, r_i, d)$ to deliver a packet from s to d along the path (s, r_i, d) is:

$$\bar{n}_{s,d}(s, r_i, d) = \bar{n}_{s,r_i} + \bar{n}_{r_i,d} = \frac{1}{d_{s,r_i}} + \frac{1}{d_{r_i,d}} = \frac{1}{0.2} + \frac{1}{1} = 6 \quad \forall i. \tag{4.3}$$

On the other hand, opportunistic routing generalizes the multihop paradigm by means of *cooperative relaying* (Figure 4.3): the source treats all the available relays as a unique unit that cooperatively forwards the packet to the destination. In other words, the next hop selection is postponed at the receiver side, allowing it to take advantage of a transmission that reaches whichever neighbor. In such a way, with reference to the previous example and assuming that the link success

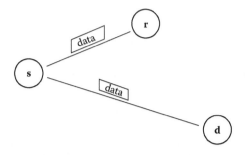

Figure 4.3   Opportunistic forwarding.

events are statistically independent, the combined link (s,r_o) has the following delivery ratio:

$$d_{s,r_o} = 1 - \prod_{i=1}^{5}(1 - d_{s,r_i}) = 1 - (1 - 0.2)^5 = 0.67 \qquad (4.4)$$

and thus the average number of link transmissions $\bar{n}_{s,d}(s,r_o,d)$ to deliver a packet from s to d using opportunistic routing is:

$$\bar{n}_{s,d}(s,r_o,d) = \bar{n}_{s,r_o} + \bar{n}_{r_o,d} = \frac{1}{d_{s,r_o}} + \frac{1}{d_{r_o,d}} = \frac{1}{0.67} + \frac{1}{1} = 2.49, \qquad (4.5)$$

which allows opportunistic routing to achieve 2.4 times the traditional routing throughput.

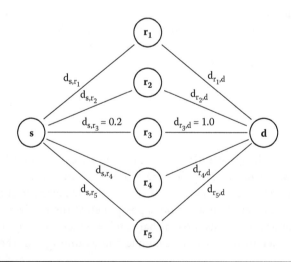

Figure 4.4   Diamond topology.

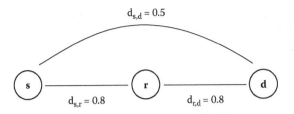

$$d_{s,d} = 0.5$$

$$d_{s,r} = 0.8 \qquad d_{r,d} = 0.8$$

Figure 4.5   Linear topology.

4.2.1.1.1 Route Opportunism

Traditional ad hoc routing tries to fortify the wireless channel so that it behaves like a wired channel by selecting links with the highest delivery ratios [10]. This choice often involves a trade-off between link quality and routing progress.

Let us consider the linear topology shown in Figure 4.5 where node s sends a packet to d along one of the possible paths $\{(s,d); (s,r,d)\}$. Traditional routing singles out the next hop at the sender side. So, if r is chosen as the next hop, the link quality is good and no retransmission is required with probability $d_{s,r} = 0.8$, but the routing progress is small. Alternatively, if the final destination is chosen as the next hop, the highest routing progress is achieved if the packet reaches the destination; however, since the link quality is poor, the probability of single transmission is just $d_{s,d} = 0.5$. The average number of link transmissions $\bar{n}_{s,d}$ to deliver a packet from s to d depends, of course, on the routing strategy—that is, it depends on the selected route:

$$\bar{n}_{s,d}(s,r,d) = \bar{n}_{s,r} + \bar{n}_{r,d} = \frac{1}{d_{s,r}} + \frac{1}{d_{r,d}} = \frac{1}{0.8} + \frac{1}{0.8} = 2.50$$

$$\bar{n}_{s,d}(s,d) = \bar{n}_{s,d} = \frac{1}{d_{s,d}} = \frac{1}{0.5} = 2.0 \tag{4.6}$$

In contrast, in opportunistic routing the sender broadcasts the packet, allowing it to pick as a relay the node closest to the destination among the nodes that receive the packet. In this way, it is able to opportunistically leverage unexpected paths related to node mobility and/or changes in wireless propagation conditions; in other words, it exploits *route opportunism*.

With reference to the previous example, let us denote with $e_{i,j}$ the event "node j is the closest node to the destination among those that have received the packet sent by i" and denote with $\bar{e}_{i,j}$ the event "no one node has received the packet sent by i." Clearly, the event $e_{i,j}$ represents the amount of progress toward the destination reached by the packet.

At the first transmission, we have three possible mutually exclusive events: $e_{s,d}$, $e_{s,r}$ and $e_{s,s}$, and the related probabilities are:

$$p_{s,d} = P(e_{s,d}) = d_{s,d} = \frac{1}{2}$$

$$p_{s,r} = P(e_{s,r}) = (1 - d_{s,d})d_{s,r} = \frac{1}{2}\frac{4}{5}$$

$$p_{s,s} = P(e_{s,s}) = (1 - d_{s,d})(1 - d_{s,r}) = \frac{1}{2}\frac{1}{5}. \tag{4.7}$$

If $e_{s,d}$ happens, the packet has reached the destination and no additional transmissions are required. Otherwise, a second transmission is needed, and if $e_{s,r}$ occurs, the possible events are $e_{r,d}$ and $e_{r,r}$ with probabilities respectively:

$$p_{r,d} = P(e_{r,d}) = d_{r,d} = \frac{4}{5}$$

$$p_{r,r} = P(e_{r,r}) = 1 - d_{r,d} = \frac{1}{5}. \tag{4.8}$$

In contrast, if $e_{s,s}$ happens, the events and the probabilities are the same as the first transmission, Equation (4.7). Further transmissions follow the same event flow, as shown in Figure 4.6, where the number of links from the root to a leaf accounts for the number of transmissions.

By exploring all the branches of the event flow, after simple algebraic manipulations, the average number of opportunistic link transmissions $\bar{n}_{s,d}(s, n_o, d)$ for the considered topology is:

$$\bar{n}_{s,d}(s, n_o, d) = \sum_{i=1}^{\infty} (p_{s,s})^{i-1} (ip_{s,d} + p_{s,r} \sum_{j=1}^{\infty} (i+j) p_{r,d} (p_{r,r})^{j-1}) =$$

$$= \frac{1}{1 - p_{s,s}} \left[\frac{p_{s,d}}{1 - p_{s,s}} + \frac{p_{s,r} p_{r,d}}{1 - p_{r,r}} \left(\frac{1}{1 - p_{s,s}} + \frac{1}{1 - p_{r,r}} \right) \right] =$$

$$= 1.8381. \tag{4.9}$$

Therefore, opportunistic routing outperforms traditional routing, and it can be shown that the throughput gain increases with the number of links exploited by the routing procedure [3].

We note that the selection of the next hop at the receiver side is the distinguishing feature that differentiates opportunistic routing from multipath routing

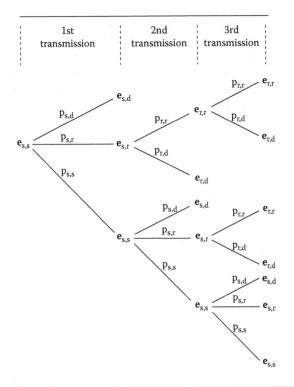

Figure 4.6 Opportunistic routing flow.

[6,21,24,31,32]. In fact, both exploit the spatial diversity (i.e., the availability of multiple routes to increase the throughput and to gain resilience against unreliable links), but since multipath routing singles out the next hop at the sender side, it exploits only a subset of the opportunities offered by the wireless propagation.

4.2.2 Challenges

The major challenge in opportunistic routing is to maximize the routing progress of each data transmission toward the destination without causing duplicate transmissions or incurring significant coordination overhead.

In order to achieve the potential benefits of opportunistic routing and avoid the above-mentioned problems, an effective protocol should implement the following tasks according to a distribute strategy:

1. candidate selection
2. forwarder election
3. forwarding responsibility transfer
4. duplicate transmission avoidance

Candidate selection guarantees that among all the neighbor nodes, only those closer to the destination than the actual forwarder can potentially become the next hop. In fact, it is completely useless to send a packet toward nodes farther from the destination than the actual forwarder, since in this case no routing process would be achieved at all. We note that the more accurate the forwarder election, the less coordination overhead is required for the responsibility transfer phase.

The *forwarder election* provides a mechanism to single out, among all the candidates that have successfully received the packet, the one that is closest to the destination. In other words, the forwarder election allows the selection of the next hop at the receiver side. Clearly, the more accurate the forwarder election is, the more the throughput increases.

The *forwarding responsibility transfer* allows the nodes involved in the forwarding process—the actual forwarder plus the candidates—to become aware of the winner of the election. The responsibility transfer is the distinguishing feature that differentiates opportunistic routing from flooding. In fact, in both opportunistic routing and flooding, multiple nodes receive the packet. However, unlike flooding, opportunistic routing allows only one node at a time to be in charge of packet forwarding. The more effective the responsibility transfer is, the less duplicate transmissions happen, and thus, less overhead is generated by the duplicate transmission avoidance mechanism.

Finally, *duplicate transmission avoidance* is required only in cases of imperfect responsibility transfer. If the forwarding responsibility is correctly transferred to the winning forwarder, there is only one node in charge of packet forwarding at any one time. In contrast, several packet transmissions occur but only one is innovative—i.e., the one made by the winning forwarder. In such a case, a mechanism is necessary to stop useless transmissions, and the more the mechanism is effective, the less network throughput is wasted.

Figures 4.7 through 4.10 present an example of the different tasks in which a node *f* has to forward a packet toward *d,* and it has 5 neighbors around it, namely

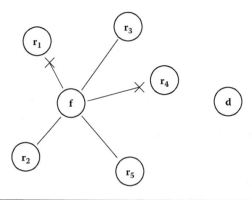

Figure 4.7   Broadcast packet forwarding.

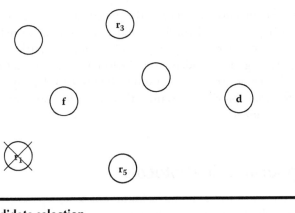

Figure 4.8   Candidate selection.

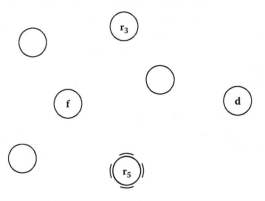

Figure 4.9   Forwarder election.

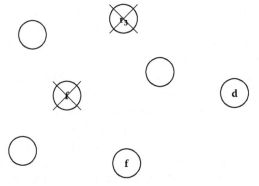

Figure 4.10   Responsibility transfer.

r_1–r_5. First, the forwarder broadcasts the packet, which is received only by nodes r_1, r_3, and r_5 (Figure 4.7). Then, the candidate selection task singles out as possible forwarders only nodes r_3 and r_5 (Figure 4.8), since r_1 is farther from the destination than f. Among the candidates, the closest node is r_5, which wins the forwarder election and becomes the next forwarder (Figure 4.9). Finally, the responsibility transfer informs the previous forwarder along with the losing candidate that r_5 won the election (Figure 4.10).

4.3 Opportunistic Protocols

In this section we describe the main features of some representative opportunistic routing protocols. The Extremely Opportunistic Routing (ExOR) (Section 4.3.1) is the most popular opportunistic routing protocol, while both MORE (Section 4.3.3) and MIXIT (Section 4.3.4) generalize the opportunistic routing paradigm by adopting, respectively, packet-level and symbol-level network coding. Simple Opportunistic Adaptive Routing (SOAR) (Section 4.3.2) proposes a simple packet-level responsibility transfer process based on time division multiple access (TDMA). The same mechanism is adopted by Opportunistic DHT-based Routing (ODR) (Section 4.3.6), but it is the first protocol to propose a scalable mechanism to distribute loss rate estimates across the network. Finally, the Multi-Channel Extremely Opportunistic Routing (MCExOR) protocol (Section 4.3.5) extends opportunistic routing to multichannel environments.

In Section 4.3.7 the main characteristics of the considered protocols are summarized, and a qualitative comparison is offered covering the advantages and drawbacks of each.

4.3.1 Extremely Opportunistic Routing

Extremely Opportunistic Routing (ExOR) [1,2] is the most popular opportunistic routing protocol and one of the first protocols proposed to exploit the broadcast nature of wireless communications for increasing resilience and throughput.

ExOR assumes that the estimates of the path loss rates for each pair of nodes are available at each node. Such loss rates are evaluated by means of a metric similar to that of Expected Transmission Count (ETX) [10]. Although the authors suggest using a link-state flooding technique to distribute loss rate estimates across the networks, in the performance evaluation they do not account for it by resorting to a simple centralized mechanism for loss rate distribution.

To contain the overhead due to the forwarding responsibility transfer mechanism, ExOR operates on batches of packets; that is, the receiving nodes buffer the packets until the end of the batch. Clearly this increases the end-to-end delay and makes ExOR unsuitable for real-time applications. Moreover, the authors point out that the batches could badly interact with the TCP congestion avoidance mechanism, since, in the presence of low loss rates, the window's size would limit the batch sizes.

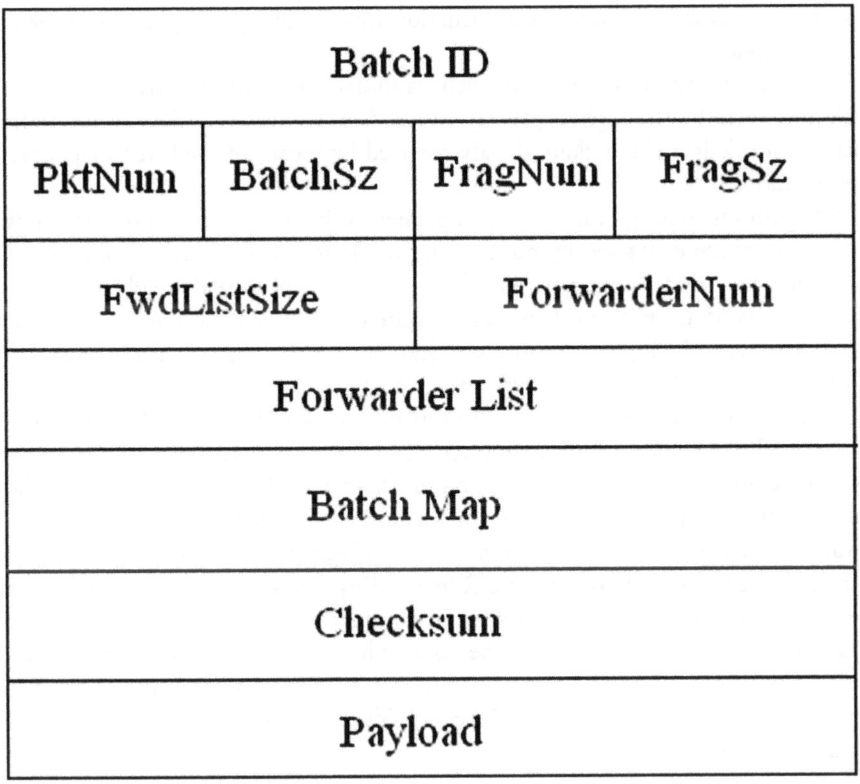

Batch ID			
PktNum	BatchSz	FragNum	FragSz
FwdListSize		ForwarderNum	
Forwarder List			
Batch Map			
Checksum			
Payload			

Figure 4.11   ExOR packet header.

The loss rates are used for both candidate selection and the forwarder election. According to ExOR, the sender must include in the header of each packet the list of candidates (Figure 4.11), namely the *forwarder list*, prioritized by closeness to the destination according to the ETX-like metric. For a given batch, the forwarder list never changes. Thus, both the candidate set and the forward election are predetermined at the sender side during the transmission of the first packet of the batch. Clearly, this could potentially reduce the opportunism of the protocol.

The forwarder responsibility transfer mechanism implements an implicit strategy based on the *batch map* field in the packet header (Figure 4.11). This field lists, for each packet in the batch, the sender's best guess of the highest priority node that has received such a packet. From an operational point of view, when a node receives a packet it first checks if it is included in the forwarder list. If so, it first buffers the packet and then updates its local batch map by replacing an entry if the packet's header indicates a higher-priority node. If the header does not include a higher-priority node, it simply discards the packet. The batch map acts like a gossip mechanism, carrying reception information from higher-priority nodes to lower-priority nodes.

When the batch is complete, each candidate forwards the packets not yet acknowledged by the highest priority candidates. Clearly, each forwarded packet also acknowledges the packets already received by means of the batch map stored in its header.

The timing among candidates is implemented by means of local timers, whose expire times are estimated by nodes using the header fields. Each candidate first estimates the sender's transmission rate using an exponential weighted moving average (EWMA) filter and then uses that rate to estimate the batch end time. The candidate with the highest priority will start forwarding packets at the batch end time. It will also delay its transmission according to its priority.

Such a TDMA strategy avoids collisions among candidates, since ExOR exploits marginal links where carrier sense often does not operate satisfactorily. However, it introduces considerable overhead, and for this reason ExOR operates on batches of packets and the candidates do not forward any packet when the batch map indicates that over 90% of the batch has been received by higher-priority nodes. Moreover, a major drawback of ExOR forwarding responsibility transfer is that its overhead is proportional to the number of candidates. For this reason, ExOR limits the candidate set size by selecting the nodes whose ETX metric does not exceed a certain threshold. Another issue related to this TDMA strategy is that it prevents the candidates from exploiting spatial bandwidth reuse by allowing only one transmission at a time.

ExOR duplicate transmission avoidance is a passive distribute procedure based on the gossip mechanism implemented by the batch lists. Since there is no explicit cancellation of redundant transmissions, a candidate can need several responsibility transfer phases to become aware that its buffered packets are not innovative.

4.3.2 Simple Opportunistic Adaptive Routing

The Simple Opportunistic Adaptive Routing (SOAR) protocol [29] tries to solve one of the issues of ExOR, the lack of support for multiple simultaneous flows due to batch processing, by introducing an explicit forwarding responsibility transfer.

Like ExOR, SOAR implements a predetermined candidate selection process based on the estimates of the path loss rates for each pair of nodes according to the ETX metric. The candidate set, namely the *forwarder list*, is included in the packet header and is prioritized by closeness to the destination.

When a candidate receives a packet, it stores the packet in a buffer and sets a timer based on its priority (i.e., its position in the forwarder list). The higher the candidate priority is, the earlier the timer will expire. Since the node broadcasts the packet, when a timer expires the other candidates can become aware that a node closest to the destination is in charge of packet forwarding and will therefore discard the packet.

Thus, like ExOR, both the candidate selection and the forwarder election processes are predetermined at the sender side on the basis of the loss rates. However, unlike ExOR, the SOAR forwarder responsibility transfer process implements an explicit acknowledgment strategy based on the packet reception. Moreover, while ExOR implements a batch-level acknowledgment, SOAR adopts a packet-level acknowledgment. That is, each candidate becomes aware of the winner of the election of each packet.

Clearly, the priority-based timers require that all the candidates can hear each other. To ensure this condition, SOAR selects the allowed candidates at the sender side in order to avoid diverging routes. The candidate selection consists of two phases: (1) shortest-path candidate selection—that is, the selection of the nodes belonging to the shortest-path, and (2) near shortest-path candidate selection— that is, the selection of additional nodes that allow an increase in opportunities but at the same time do not produce diverging routes.

Assume that node i belongs to the shortest route between the source and the destination (i.e., its distance to the destination is the shortest one). Then i will select neighbor j as a candidate if all of the following conditions hold:

1. j is closer (according to the ETX metric) than i to the destination
2. the quality of the link between i and j is above a certain threshold
3. the qualities of the links between j and each other candidate are above the threshold

The first constraint ensures routing progress, while the second and the third conditions assure that the actual forwarder plus the candidate set are connected to avoid duplicate transmissions. However, these constraints do not avoid the presence of diverging routes if they are used by nodes that do not belong to the shortest path.

As an example, let us consider the topology depicted in Figure 4.12, where node s wants to send a packet to node d. According to the previous constraints, s selects as

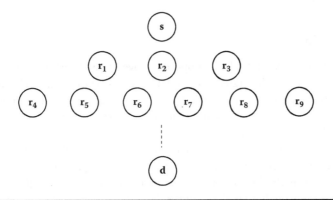

Figure 4.12 SOAR example.

candidates the nodes r_1, r_2, and r_3, r_1 selects the candidates r_4, r_5, and r_6, and r_3 selects the candidates r_7, r_8, and r_9. If it happens that r_1 and r_3 do not hear each other forwarding the packet due to some packet loss, they each forward a copy of the packet. Since r_4 is far away from r_9, these two nodes further perform duplicate forwarding and the paths will further diverge and yield many duplicate transmissions.

For these reasons, nodes that do not belong to the shortest path use the following additional constraints for candidate selection. Assuming that node i does not belong to the shortest path, i first determines the node belonging to the shortest path that is closest to the destination, say node j. Then j becomes a candidate for i if it is closer than i is to the destination. Moreover, for each node k in j's forwarding list, i adds k in its candidate set if:

■ k is closer than i to the destination
■ the quality of the link between i and k is above a certain threshold

By applying the above constraints to the example reported in Figure 4.12, we observe that even if r_1 and r_3 perform duplicate forwarding, since their forwarding lists include only r_6 and r_7, the routes do not further diverge.

Besides the implicit duplicate transmission avoidance based on the diverging route prevention, SOAR also implements an explicit mechanism based on selective and piggybacked acknowledgments (ACKs). The ACKs are selective, since the same ACK can acknowledge multiple data packets, and they are piggybacked because, if there is a data packet in the queue, the acknowledgment is stored in the data packet header, limiting the throughput related to the duplicate transmission avoidance.

We note that SOAR shares some similarities with the MAC-layer anycast mechanism proposed in [14], where a sender sends a request-to-send (RTS) packet, and multiple receivers respond to the RTS according to their closeness to the destination. However, in [14] the reception of the RTS does not guarantee the reception of the subsequent data packet.

4.3.3 MORE

The MORE protocol [8] has been proposed to overcome the issues related to ExOR forwarding responsibility transfer, mainly the lack of spatial reuse. The key feature of MORE is the adoption of network coding at the packet level (intra-flow), and in the following we provide two examples to show the synergy between opportunistic routing and network coding.

In the first example, we consider the linear topology shown in Figure 4.13, where node s has to send two packets to node d, namely p_1 and p_2. While traditional ad hoc routing sends the packets unicast along one of the paths $\{(s, d); (s, r, d)\}$ by selecting the next hop at the sender side, opportunistic routing simply broadcasts the packets and the next hop selection happens at the receiver side. However,

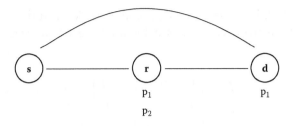

Figure 4.13   MORE unicast example.

opportunistic routing requires a certain amount of coordination, which introduces additional overhead. In fact, if we suppose that r receives both the packets but only one has been overheard by the destination, r has no way to guess which packet it has to forward.

MORE exploits the network coding to solve such an issue: r simply sends a random linear combination of the received packets p_1 and p_2—the sum $p_1 + p_2$—and the destination will retrieve the missing packet without any additional coordination. In other words, MORE adopts the network coding to accomplish the *forwarder responsibility transfer*.

The second example (Figure 4.14) illustrates a multicast transmission: the source s has to multicast four packets—p_1 through p_4—to three nodes, r_1 through r_3. We assume that each node receives the packets shown in the figure. Without network coding, the source has to retransmit all four packets. However, with network coding, it is sufficient to transmit two linear combinations of the four packets, which will be used by the destination to retrieve the original packets. For example, if the sender sends:

$$p_1' = p_1 - p_2 + p_3 - p_4$$

$$p_2' = p_1 + 2p_2 + p_3 + 3p_4$$

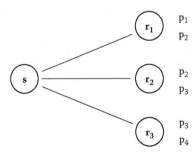

Figure 4.14   MORE multicast example.

the node r_1, which has received p_1, p_2, p_1' and p_2', retrieves all the original packets by inverting the matrix of coefficients and multiplying it with the received packets, as follows:

$$\begin{pmatrix} p_1 \\ p_2 \\ p_3 \\ p_4 \end{pmatrix} = \begin{pmatrix} 1 & 0 & 0 & 0 \\ 0 & 1 & 0 & 0 \\ 1 & -1 & 1 & -1 \\ 1 & 2 & 1 & 3 \end{pmatrix}^{-1} \begin{pmatrix} p_1 \\ p_2 \\ p_1' \\ p_2' \end{pmatrix} \qquad (4.10)$$

which reduces the number of retransmissions from four packets to just two.

MORE shares several features with ExOR. Both protocols implement a predetermined candidate selection process based on the estimates of the path loss rates for each pair of nodes, and both adopt the ETX metric [10] to estimate such loss rates. Both include the *forwarder list* in the packet header, prioritized by closeness to the destination, and both operate on batches of packets. Finally, both limit the candidate set size to reduce the overhead.

However, unlike ExOR, the forwarder election process allows multiple nodes to forward the packets. In fact, when a node receives a packet, it first checks whether it is in the packet's forwarder list. If so, the node checks if the packet is an *innovative* one; that is, whether it is linearly independent of the packets of the same batch previously received. If both conditions are satisfied, the node stores the packet in the buffer and broadcasts a linear combination of the received packets.

Thus, MORE does not implement any forwarder election within the candidate set, and the forwarding responsibility transfer is implicit (i.e., it is based on the packet reception event). As a distinguishing feature of MORE, the classical CSMA/CA strategy offered by the 802.11 MAC layer is used to avoid collisions among forwarder nodes.

Another difference between MORE and ExOR is that each packet sent by MORE is a coded packet—i.e., a linear combination of all the packets in the batch. Therefore, a duplicate transmission occurs every time a packet is linear dependent from the previously received ones. MORE does not use any explicit strategy to avoid duplicate transmissions, since there is no explicit cancellation of redundant transmissions. Instead, it resorts to the path loss rates to estimate the number of transmissions needed to forward a packet to a node closest to the destination, and such estimates implicitly limit duplicate transmission events. Each time that a packet is received from the farthest node, a *credit counter* is incremented by such an estimate, and each time that the node forwards a packet, its credit counter is decremented by one.

An explicit acknowledgment strategy is used to notify the source that a batch is correctly received by the destination, and the ACK is routed using traditional unicast routing. Clearly, batch size affects the MORE overhead because the smaller the batch sizes, the more frequent the ACKs. Moreover, the batch size also affects the duplicate transmission occurrence because the smaller the batch, the more likely the duplicate transmission event.

4.3.4 MIXIT

The MIXIT [18,19] protocol extends the network coding proposed by MORE at the symbol level with three differences. First, MIXIT deals with packets with errors, while MORE does not. Second, MIXIT network coding is an end-to-end rateless error-correcting code while MORE network code cannot correct errors. Third, MIXIT designs a MAC that exploits looser constraints on packet delivery to significantly increase concurrent transmissions, while MORE carrier sense requires correct packet delivery, preventing it from achieving high concurrency.

The key insight MIXIT is that, by insisting on receiving fully correct packets, traditional protocols are missing the bulk of their opportunities. In fact, over long links it is hard to receive the whole packet correctly. On the other hand, it is likely that each symbol will be received correctly by some node thanks to spatial diversity [25]. MIXIT opportunistically exploits such partial packets received at the intermediate nodes to assemble them into a complete packet at the destination, thus increasing the network throughput.

The assumption behind MIXIT is the availability at the physical layer of a confidence measure for each decode symbol [15,33]. This allows the nodes to identify which symbols in a corrupt packet are likely correct and to forward them. More specifically, a symbol is error free if its confidence value is above a threshold, γ, and faulty otherwise, since as γ increases the probability that a symbol is corrupted becomes vanishingly small [33].

Such an assumption is used to define a symbol-level network coding that also works as a rateless error-correcting code, addressing one of the main challenges in opportunistic routing: duplicate transmission avoidance. In fact, with network coding, nodes forward random linear combinations of their correctly received symbols, reducing the probability of duplicate transmission. Moreover, since MIXIT exploits symbol-level network coding by forwarding symbols belonging to corrupt packets, there is a chance that a forwarded symbol is incorrect. Therefore, duplicate transmissions provide an amount of redundancy to correct corrupted symbols; in other words, they behave as a rateless error-correcting code. To provide an example of how MIXIT works, let us consider the scenario in Figure 4.15, where a source s wants to deliver two packets, namely p_a and p_b, to the destination d, and where there are two possible relays, r_1 and r_2. We assume that when the source broadcasts p_a and p_b, the nodes in the network receive some corrupted symbols, and in particular the relays receive less corrupted symbols than the destination. Figure 4.15 illustrates such corrupted symbols using grey cells. Due to spatial diversity [25], however, the few corrupted symbols received by r_1 and r_2 are unlikely to be in the same locations.

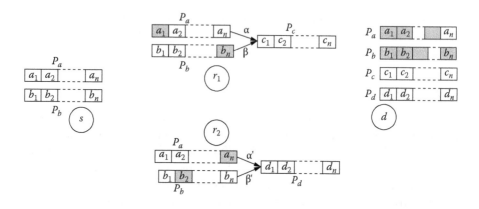

Figure 4.15 MIXIT example.

Since the faulty symbols can be recognized thanks to the confidence measure, according to MIXIT the nodes forward linear combinations of the received error-free symbols. In particular, if we assume that a_i and b_i are the i-th correct symbols in p_a and p_b, respectively, the node r_1 picks two random numbers α and β and creates a coded packet p_c where the i-th symbol, c_i, is computed as follows:

$$c_i = \begin{cases} \alpha a_i + \beta b_i & \text{if } a_i \text{ and } b_i \text{ are clean symbols} \\ \alpha a_i & \text{if } a_i \text{ is clean and } b_i \text{ is faulty} \\ \beta b_i & \text{if } a_i \text{ is faulty and } b_i \text{ is clean} \\ \text{none} & \text{if } a_i \text{ and } b_i \text{ are both faulty} \end{cases} \qquad (4.11)$$

Similarly, r_2 generates a coded packet p_d by picking two random values α' and β' and applying the same logic in the above equation.

When r_1 and r_2 broadcast their respective packets, p_c and p_d, the destination receives corrupted versions where some symbols are incorrect as shown in Figure 4.15. Thus the destination has four partially corrupted receptions: p_a and p_b, directly overheard from the source and containing many erroneous symbols, and p_c and p_d, which contain a few erroneous symbols. For each symbol at the i-th position, the destination needs to decode two original symbols a_i and b_i. As long as the destination receives two uncorrupted independent symbols in location i, it will be able to perform the decoding [13]. For example, with respect to the second symbol, the destination has received:

$$c_2 = \alpha a_2 + \beta b_2$$
$$d_2 = \alpha' a_2 \qquad (4.12)$$

Given that the header of a coded packet contains the multipliers (e.g., α and β), the destination has two linear equations with two unknowns, a_2 and b_2, which are easily solvable. Once the destination has decoded all symbols correctly, it broadcasts an ACK, causing the nodes to stop forwarding packets.

Apart from the symbol-level network coding, MIXIT behaves identically to MORE. Therefore, both share the same advantages and the same issues. We note also that besides MORE and MIXIT, other recently proposed protocols deal with network coding in opportunistic routing [22,34].

4.3.5 Multi-Channel Extremely Opportunistic Routing

The Multi-Channel Extremely Opportunistic Routing (MCExOR) protocol [39] extends the ExOR protocol by adopting multi-channel forwarding requiring a single RF transceiver per device. The simultaneous use of multiple RF channels is in fact a promising approach to increase the capacity of multihop wireless networks, and MCExOR improves the network performance by choosing the RF channel with the most promising candidate set for every transmission. However, the multi-channel approach introduces new issues related to channel management, and MCExOR does not deal with these issues because it assumes that the channel assignment is decoupled by the routing protocol.

While both candidate selection and the forward election processes are simple tasks in ExOR, MCExOR introduces the additional issue related to choosing the transmission channel. Such an issue involves the construction of a candidate set for each RF channel, and then the selection of the most promising candidate set.

The selection of the candidates for each channel is based on the ETX metric, which is very similar to the ExOR metric. In contrast, the selection of the most promising candidate set requires a moderate amount of sophistication.

Using $p_1(x, y)$ to define the success probability of the link between nodes x and y and with $g(x, y, z)$ the ETX metric for the path between source x and destination z using y as the next hop, the priority q of a candidate set $C_s = \{c_i\}_{i=1}^n$ for the forwarder w is heuristically defined as:

$$q(C_s) = \sum_{i=1}^{n} g(w, c_i, d) \frac{p_c(w, c_i)}{p_{nc}(w)}$$ (4.13)

where d is the destination, $p_c(w, c_i)$ is the probability that the i-th candidate is w's next forwarder—that is, the probability that the packet sent by w is not received by the $i - 1$ candidates with highest priority:

$$p_c(w, c_i) = p_l(w, c_i) \prod_{j=1}^{i-1} (1 - p_l(w, c_j))$$ (4.14)

and $p_{nc}(w)$ is the probability that the packet is not received by any of the candidates:

$$p_{nc}(w,c_i) = \prod_{j=1}^{n} (1 - p_l(w,c_j)). \qquad (4.15)$$

Like SOAR, the MCExOR forward responsibility transfer is based on a TDMA strategy. However, MCExOR uses the TDMA mechanism to regulate the node access for ACK packet transmission. From an operational point of view, when a candidate receives a packet, it stores the packet in a buffer and sets a timer based on its priority (i.e., its position in the forwarder list). The higher the candidate priority, the sooner the timer will expire. When a timer expires, the node sends an ACK packet to the source, and the other candidates send their ACKs according to their priorities. In such a way, all the candidates can become aware of which node is in charge of packet forwarding.

4.3.6 Opportunistic DHT-Based Routing

The Opportunistic DHT-based Routing (ODR) protocol [7] resorts to a location-aware addressing schema [4,5,11], which allows it to group nodes based on their addresses. This approach lets nodes estimate, by means of the ETX metric, their distances from sets of nodes sharing the same address prefix, instead of individually tracking each node. In this way, ODR provides a scalable strategy for loss rate distribution, which is a common issue of opportunistic routing protocols.

However, such a procedure requires the availability of a distribute procedure to allow nodes to retrieve the destination addresses before starting a communication. ODR accomplishes this task by resorting to a Distributed Hash Table (DHT) system, which exploits a globally known hash function $h(\cdot)$, defined on the IP address space and with values in the location-aware address space.

Every node is part of the DHT system, storing a subset of *pairs* <IP address, location-dependent address> in accordance with the hash function. Specifically, the pair $<ip_1, add_1>$ is stored by the node whose location-dependent address is equal to $h(ip_1)$, namely the *rendezvous node*. Thus, to find a location-dependent address, a node simply sends a pair request to the rendezvous node, as shown in Figure 4.16. After the reception of the pair reply, the node is able to establish the communication.

More in detail, Opportunistic DHT-based Routing (ODR) assigns location-dependent addresses, namely strings of l bits, to nodes by means of a distribute procedure that resorts to locally broadcasted hello packets. The address allocation procedure guarantees that nodes sharing a common address prefix are close in the physical topology, allowing the routing tables to easily group nodes.

ODR represents the address space as a *complete binary tree* of $l + 1$ levels—that is, as a binary tree in which every vertex has zero or two children and all leaves are at the same level (Figure 4.18a). In the tree structure, each leaf is associated with an address, and an inner vertex of level k, namely a *level-k subtree*, represents a set

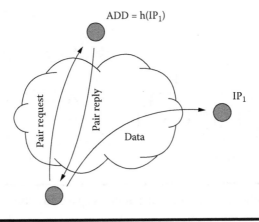

Figure 4.16 Location-dependent address discovery.

of leaves (that is, a set of peer identifiers) sharing a prefix of $l - k$ bits. For example, with reference to Figure 4.18a, the vertex with the label *01X* is a level-1 subtree and represents the leaves *010* and *011*.

Defining a *level-k sibling* of a leaf as the level-*k* subtree, which shares the same parent with the level-*k* subtree to which the leaf belongs, and referring to the previous example, the vertex with the label *1XX* is the level-2 sibling of the address *000*.

By means of the sibling concept, nodes can reduce the overhead due to maintaining a distance state by a logarithm factor. Each node stores a limited-size distance table composed of *l* entries, one for each sibling, and the *k*-th section contains the distance estimated according to the ETX metric with the *nearest* node whose location-dependent address belongs to the level -*k* sibling. As shown in Figure 4.17, each node stores a limited-size distance table.

Clearly, this approach raises a new problem because the hierarchy related to the sibling concept gives rise to an estimate inaccuracy. In fact, the *k*-th section stores the estimated distance toward the nearest node whose address belongs to the level-*k* sibling; in other words, the section stores a lower bound on the distance.

The proposed solution to this issue requires that when a node forwards a packet, it stores its location-dependent address in the packet header along with its estimated

Destination	Path quality	Route log
011	1.60	001
00X	3.80	010
1XX	1.25	010

Figure 4.17 Node 010 routing table.

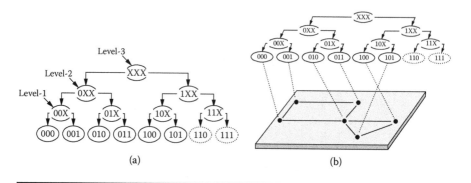

Figure 4.18 Relationship between the address space structure and the physical topology.

distance from the destination. A receiving node determines whether its overlay distance to the destination (i.e., the length of the address prefix shared by the node address and the destination address) is shorter than the forwarding overlay distance and then checks whether its path quality is better than the quality of the forwarder node. If both of these checks fail, the node does not belong to the candidate set and it stores the packet in its ACK queue. If both checks are successful, it stores the packet in its packet queue together with a delay time evaluated according to the following heuristic relation:

$$delay = \tau * \frac{q_p(r,d)}{q_p(f,d)} * \left[\frac{1}{o_d(r,d) - o_d(f,d) + 1} \right] \qquad (4.16)$$

where τ is the maximum delay time (2 seconds in our implementation), f is the forwarding node, r is the receiving node, d is the destination node, q_p is the estimated quality, and o_d is the overlay distance. By means of this heuristic approach for the delay estimation, the authors account for the estimated inaccuracy mentioned before, since the ratio between the estimated qualities ratio is weighted by a factor (i.e., the term in the square brackets in Equation [4.16] depending on the overlay distances) that measures the size of the cluster of nodes, namely the siblings, to which the qualities refer.

Thus, the delay times allow nodes to implement a distributed candidate election procedure by exploiting a TDMA-based scheduling. Because the closest node stores its estimated distance from the destination in the packet header, and since it is the first node that forwards the packet, the other candidates can listen to the packet transmission and therefore give up responsibility for packet forwarding.

Such a strategy does not require explicit acknowledgment for forwarding responsibility transfer, although it is not tolerant of the hidden terminal problem. In such a case, ODR resorts to explicit acknowledgment.

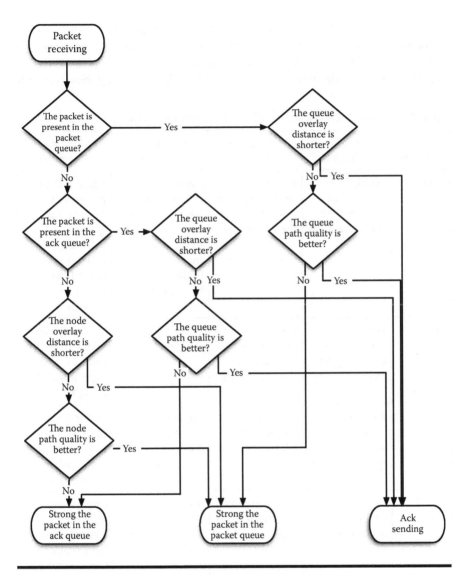

Figure 4.19   Packet forwarding process.

Figure 4.19 gives a detailed description of the whole forwarding process in a flow chart.

4.3.7 *Comparison of the Considered Protocols*

In this section we summarize the main characteristics of the considered protocols and offer a qualitative comparison in terms of advantages and drawbacks.

More specifically, Table 4.1 compares the strategies adopted by the different protocols for each task, which are *candidate selection, forwarder election, responsibility transfer*, and *duplicate transmission avoidance*. With respect to candidate election, all the considered protocols adopt an ETX-based strategy. However, only

Table 4.1   Basic Characteristics of the Considered Protocols

Protocol	Candidate Selection	Forwarder Election	Responsibility Transfer	Duplicate Tx Avoidance
ExOR	Fixed at the sender side according to the ETX metric	Established at the sender side according to the ETX metric	Implicit at the batch level based on TDMA and according to the gossip mechanism implemented by the batch maps	Implicit at the packet level based on the gossip mechanism, explicit at the batch level based on ACK packets
SOAR	Fixed at the sender side according to the ETX metric	Established at the sender side according to the ETX metric	Explicit based on TDMA	Implicit based on the diverging route avoidance, explicit based on ACKs
MORE	Fixed at the sender side according to the ETX metric	None: multiple forwarder allowed	Implicit at the batch level based on packet reception	Implicit based on the estimation of the number of transmissions
MIXIT	Fixed at the sender side according to the ETX metric	None: multiple forwarder allowed	Implicit at the packet level based on packet reception	Implicit based on the estimation of the number of transmissions
MCExOR	Fixed at the sender side according to the ETX metric	Established at the sender side according to the ETX metric	Explicit based on TDMA	Explicit based on ACKs
ODR	Dynamic at the receiver side according to the ETX metric	Dynamic at the receiver side according to the ETX metric	Explicit based on TDMA	Implicit based on packet reception and explicit based on ACKs

Table 4.2   Comparison of the Considered Protocols

Protocol	Advantage	Drawbacks
ExOR	Low overhead due to explicit acknowledgment	Complex forward responsibility transfer likely to produce duplicate transmissions; operates on batch of packets
SOAR	Simple forward responsibility transfer operates on single packets	Limited opportunities due to the route divergence avoidance
MORE	Multiple forwarders allowed	Complicated duplicate transmission avoidance; operates on batch of packets
MIXIT	Multiple forwarders allowed	More coordination needed; operates on faulty packets
MCExOR	Simple forward responsibility transfer increased throughput thanks to multi-channel	High overhead due to ACKs channel assignment
ODR	Simple forward responsibility transfer scalable mechanism for path losses distribution	Unsuitable in scenarios characterized by high mobility

ODR allows the candidate set to be dynamically selected at the receiver side, thus increasing the capability to explore the opportunities offered by wireless propagation. According to the forwarder election process, the considered protocols fall into two classes: those that allow a single forwarder and those that allow multiple forwarding by means of network coding. Clearly, the last strategy assures an increased throughput, although it operates only on batches of packets. Finally, concerning responsibility transfer and duplicate transmission avoidance, all the considered protocols adopt a TDMA-like strategy, which can or cannot take into account the candidate priority.

In Table 4.2 we synthesize the main advantages of each proposal, along with the main drawbacks.

4.4 Future Work Issues

As mentioned in Section 4.1, opportunistic routing protocols try to take advantage of the time-variant nature of the environment to provide end-to-end connectivity in scenarios where traditional networking fails.

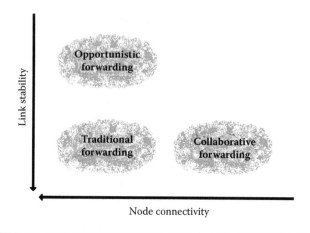

Figure 4.20   Routing protocols taxonomy.

Another class of routing protocols, the *collaborative routing* protocols, exploits a similar paradigm that assumes that the network topology is quite dense in order to assure that the presence of a persistent path between each pair of nodes is relaxed.

This class exploits the time-variant nature of the network topology to provide connectivity for sparse topologies, usually by resorting to the so-called *store-carry-forward* paradigm [17,26]. Delay-tolerant networks are the typical application domain for collaborative routing, since they aim to provide connectivity in rural and developing areas where the costs associated with traditional dense networks are not affordable.

In Figure 4.20, a taxonomy of the different classes of routing protocols is shown. The majority of routing protocols previously described relax the assumption that the wireless propagation conditions are stationary enough to allow persistent communication among neighbor nodes. The protocols belonging to the collaborative routing class relax the assumption of a dense network topology.

In the future, we expect that a new class of routing protocols able to provide connectivity when both the assumptions are not verified will be developed.

4.5 Conclusions

As has been shown in this chapter, there are a vast variety of routing protocols designed specifically for ad hoc mobile networks. These networks create a hostile routing environment due to the instability of wireless propagation conditions and the mobility of the nodes. However, with the introduction of the opportunistic paradigm, significant advances have been made toward the development of robust routing protocols that can assure end-to-end connectivity, even in hostile environments.

It is likely that there is not currently a single opportunistic routing protocol that can satisfy the needs of every conceivable network scenario. In fact, some protocols limit the set of candidates to bound the overhead for forward responsibility transfer, while also limiting the opportunities offered by the network. Other protocols resort to network coding, simplifying the responsibility transfer process but operating on batches of packets. Most of the proposed protocols need estimates of the path losses but do not provide any scalable mechanism to distribute them.

The understanding gained from these first proposals can be used, in the next few years, to improve future designs of wireless routing protocols. There still remains much to do in terms of understanding, developing, and deploying a network layer for ad hoc scenarios.

Acknowledgments

The authors would like to express their thanks to the editor, Prof. Mieso Denko, for his invitation to contribute to this book, and to Dr. Angela Sara Cacciapuoti for her insightful comments and contributions to this chapter.

References

[1] Sanjit Biswas and Robert Morris. Opportunistic routing in multi-hop wireless networks. *SIGCOMM Comput. Commun. Rev.,* 34(1):69–74, 2004.

[2] Sanjit Biswas and Robert Morris. ExOR: Opportunistic multi-hop routing for wireless networks. *SIGCOMM Comput. Commun. Rev.,* 35(4):133–144, 2005.

[3] Angela Sara Cacciapuoti, Marcello Caleffi, and Luigi Paura. Analytical evaluation of data-link transmissions in opportunistic routing. In *ICUMT '09: The IEEE International Conference on Ultra Modern Telecommunications,* October 2009.

[4] Marcello Caleffi. Mobile ad hoc networks: The DHT paradigm. In *PerCom '09: The Seventh Annual IEEE International Conference on Pervasive Computing and Communications,* pages 1–2, March 2009.

[5] Marcello Caleffi, Giancarlo Ferraiuolo, and Luigi Paura. Augmented tree-based routing protocol for scalable ad hoc networks. In *MASS '07: The IEEE International Conference on Mobile Ad hoc and Sensor Systems,* pages 1–6, October 2007.

[6] Marcello Caleffi, Giancarlo Ferraiuolo, and Luigi Paura. A reliability-based framework for multi-path routing analysis in mobile ad-hoc networks. *International Journal of Communication Networks and Distributed Systems,* 1(4-5-6):507–523, 2008.

[7] Marcello Caleffi and Luigi Paura. Opportunistic routing for disruption tolerant networks. In *AINA '09: The IEEE 23rd International Conference on Advanced Information Networking and Applications,* May 2009.

[8] Szymon Chachulski, Michael Jennings, Sachin Katti, and Dina Katabi. Trading structure for randomness in wireless opportunistic routing. *SIGCOMM Comput. Commun. Rev.,* 37(4):169–180, 2007.

[9] T. Clausen, P. Jacquet, A. Laouiti, P. Muhlethaler, A. Qayyum, and L. Viennot. Optimized link state routing protocol. In *IEEE INMIC*, December 2001.

[10] Douglas S. J. De Couto, Daniel Aguayo, John Bicket, and Robert Morris. A high-throughput path metric for multi-hop wireless routing. *Wireless Networks*, 11(4):419–434, 2005.

[11] J. Eriksson, M. Faloutsos, and S. V. Krishnamurthy. Dart: Dynamic address routing for scalable ad hoc and mesh networks. *IEEE/ACM Transactions on Networking*, 15(1):119–132, 2007.

[12] Z. J. Haas. A new routing protocol for the reconfigurable wireless networks. *Universal Personal Communications Record, 1997. Conference Record, 1997 IEEE 6th International Conference on*, 2:562–566, October 1997.

[13] T. Ho, R. Koetter, M. Medard, D. R. Karger, and M. Effros. The benefits of coding over routing in a randomized setting. In *Information Theory, 2003. Proceedings. IEEE International Symposium*, page 442, June 2003.

[14] Shweta Jain and Samir R. Das. Exploiting path diversity in the link layer in wireless ad hoc networks. *Ad Hoc Networks*, 6(5):805–825, 2008.

[15] Kyle Jamieson and Hari Balakrishnan. PPR: Partial packet recovery for wireless networks. *SIGCOMM Comput. Commun. Rev.*, 37(4):409–420, 2007.

[16] D. B. Johnson and D. A. Maltz. Dynamic source routing in ad hoc wireless networks. In *Mobile Computing*, Volume 353, pages 153–181. Kluwer Academic Publishers, 1996.

[17] Philo Juang, Hidekazu Oki, Yong Wang, Margaret Martonosi, Li S. Peh, and Daniel Rubenstein. Energy-efficient computing for wildlife tracking: Design trade-offs and early experiences with ZebraNet. In *ASPLOS-X: Proceedings of the 10th International Conference on Architectural Support for Programming Languages and Operating Systems*, volume 37, pages 96–107, October 2002.

[18] Sachin Katti and Dina Katabi. Mixit: The network meets the wireless channel. In *HotNets-VI: Proceedings of the Sixth ACM Workshop on Hot Topics in Networks*, 2006.

[19] Sachin Katti, Dina Katabi, Hari Balakrishnan, and Muriel Medard. Symbol-level network coding for wireless mesh networks. *SIGCOMM Comput. Commun. Rev.* 38, 4 (Oct. 2008), 401–412.

[20] Jonghyun Kim and Stephan Bohacek. A comparison of opportunistic and deterministic forwarding in mobile multihop wireless networks. In *MobiOpp '07: Proceedings of the 1st International MobiSys Workshop on Mobile Opportunistic Networking*, pages 9–16, 2007.

[21] S. Lee and M. Gerla. Split multipath routing with maximally disjoint paths in ad hoc networks. In *ICC '01: Proceedings of the IEEE International Conference on Communications*, pages 3201–3205, 2001.

[22] Yunfeng Lin, Baochun Li, and Ben Liang. Codeor: Opportunistic routing in wireless mesh networks with segmented network coding. In *IEEE International Conference on Network Protocols, 2008. ICNP 2008*, pages 13–22, October 2008.

[23] Chun-Pong Luk, Wing-Cheong Lau, and On-Ching Yue. An analysis of opportunistic routing in wireless mesh network. In *IEEE International Conference on Communications, 2008. ICC '08*, pages 2877–2883, May 2008.

[24] Mahesh K. Marina and Samir R. Das. Ad hoc on-demand multipath distance vector routing. *SIGMOBILE Mob. Comput. Commun. Rev.*, 6(3):92–93, 2002.

[25] Allen Miu, Hari Balakrishnan, and Can Emre Koksal. Improving loss resilience with multi-radio diversity in wireless networks. In *MobiCom '05: Proceedings of the 11th Annual International Conference on Mobile Computing and Networking*, pages 16–30, 2005.

[26] Alex (Sandy) Pentland, Richard Fletcher, and Amir Hasson. Daknet: Rethinking connectivity in developing nations. *Computer,* 37(1):78–83, 2004.

[27] C. Perkins and E. Royer. Ad hoc on-demand distance vector routing. In *2nd IEEE Workshop on Mobile Computing Systems and Applications,* pages 90–100, 1999.

[28] Charles Perkins and Pravin Bhagwat. Highly dynamic destination-sequenced distance-vector routing (DSDV) for mobile computers. In *SIGCOMM '94: ACM Conference on Communications Architectures, Protocols and Applications,* pages 234–244, 1994.

[29] E. Rozner, J. Seshadri, Y. Mehta, and Lili Qiu. Simple opportunistic routing protocol for wireless mesh networks. In *2nd IEEE Workshop on Wireless Mesh Networks, 2006. WiMesh 2006,* pages 48–54, Sept. 2006.

[30] R.C. Shah, S. Wietholter, A. Wolisz, and J. M. Rabaey. When does opportunistic routing make sense? In *3rd IEEE International Conference on Pervasive Computing and Communications Workshops, 2005. PerCom 2005 Workshops,* pages 350–356, March 2005.

[31] A. Tsirigos and Z. J. Haas. Analysis of multipath routing, part 1: The effect on the packet delivery ratio. *Wireless Communications, IEEE Transactions on,* 3(1):138–146, January 2004.

[32] A. Tsirigos and Z. J. Haas. Analysis of multipath routing, part 2: Mitigation of the effects of frequently changing network topologies. *Wireless Communications, IEEE Transactions on,* 3(2):500–511, March 2004.

[33] Grace R. Woo, Pouya Kheradpour, Dawei Shen, and Dina Katabi. Beyond the bits: Cooperative packet recovery using physical layer information. In *MobiCom '07: Proceedings of the 13th Annual ACM International Conference on Mobile Computing and Networking,* pages 147–158, 2007.

[34] Yan Yan, Baoxian Zhang, H. T. Mouftah, and Jian Ma. Practical coding-aware mechanism for opportunistic routing in wireless mesh networks. In *IEEE International Conference on Communications, 2008. ICC '08,* pages 2871–2876, May 2008.

[35] Kai Zeng, Wenjing Lou, and Hongqiang Zhai. Capacity of opportunistic routing in multi-rate and multi-hop wireless networks. *IEEE Transactions on Wireless Communications,* 7(12):5118–5128, December 2008.

[36] Jian Zhang, Y. P. Chen, and I. Marsic. Network coding via opportunistic forwarding in wireless mesh networks. In *IEEE Wireless Communications and Networking Conference, 2008. WCNC 2008,* pages 1775–1780, April 2008.

[37] Rong Zheng and Chengzhi Li. How good is opportunistic routing?: A reality check under Rayleigh fading channels. In *MSWiM '08: Proceedings of the 11th International Symposium on Modeling, Analysis and Simulation of Wireless and Mobile Systems,* pages 260–267, 2008.

[38] A. Zubow, M. Kurth, and J.-P. Redlich. Considerations on forwarder selection for opportunistic protocols in wireless networks. In *14th European Wireless Conference, 2008. EW 2008.* pages 1–7, June 2008.

[39] Anatolij Zubow, Mathias Kurth, and Jens-Peter Redlich. Multi-channel opportunistic routing. *IEEE European Wireless Conference,* April 2007.

Chapter 5

Opportunistic Routing for Load Balancing and Reliable Data Dissemination in Wireless Sensor Networks

Min Chen
University of British Columbia and Hebei Polytechnic University

Wen Ji
Chinese Academy of Sciences

Xiaofei Wang
Seoul National University

Wei Cai
Seoul National University

Lingxia Liao
University of British Columbia

Contents

Advances in microelectronics and communications enable inexpensive sensors to be deployed on a large scale and in harsh environments, where sensors need to operate unattended in an autonomous manner. As sensor nodes communicate over error-prone wireless channels with battery power, reliable and energy-efficient data delivery is crucial. These characteristics of wireless sensor networks (WSNs) make the design of routing protocols challenging [1].

Many studies are done involving WSNs with respect to energy efficiency, load balancing, and reliability. However, these design goals are generally orthogonal to each other. For example, most of the load-balancing schemes are not robust to a high link-failure rate. In this chapter, the existing load-balancing schemes are classified into two categories: *local load balancing* [2,3,4] and *global load balancing* [5]. These are also referred as hop-by-hop balancing and end-to-end balancing, respectively. To evaluate the performance of load balancing, we define the "lifetime" of a WSN as the time until the first node in the WSN drains its battery power and dies.

Local load balancing is based on a *node-centric approach*, where a hello message is broadcast by each sensor node periodically during the network operation in order to notify its neighbors of its energy changes. The interval of broadcasting such a hello message provides a trade-off between control overhead and timeliness of local energy information. Local load balancing may cause a data packet to enter an "energy bottleneck" region, where the energy levels of the sensor nodes are relatively low while the sensors outside the region may still have higher remaining energy levels. Thus, some load-balancing schemes aim to find globally load-balanced paths to achieve a higher network lifetime.

In [2,5], to reduce packet losses due to frequent link failures, a forwarding node also uses alternative (backup) nodes by setting up multiple backup next-hop nodes in advance. If the primary next-hop node fails, the medium access control (MAC) layer is not able to deliver a packet to this unreachable primary node. After several retransmission attempts, the MAC layer simply drops the packet and notifies the network layer of the transmission failure. The routing protocol then selects a backup next hop and hands the same packet (stored in the cache) down to the MAC layer. If the backup next hop also dies, these retransmissions are repeated. When the node failure rate is high, trying multiple backup nodes along with data caching severely increases the delay, reduces the effective available bandwidth, and wastes energy for unnecessary transmissions. Nevertheless, the above operation is widely used in traditional routing protocols for ad hoc and sensor networks [2,5], which operate in the following two-step manner: (1) select the next-hop node first based on a neighbor information table (denoted by NIT), as shown in Figure 5.1, and (2) forward the packet to the selected node until a predetermined number of transmissions fail. We call this a *transmitter-oriented approach.*

In this chapter, a novel opportunistic routing protocol is proposed, which facilitates load balancing and reliable data dissemination in wireless sensor networks. Compared to the traditional transmitter-oriented approach, a receiver-oriented approach is exploited to achieve both load balancing and reliability for large-scale WSNs. Thus, the proposed opportunistic routing scheme is called the load-balancing reliable routing protocol (RLRR) [7]. In the receiver-oriented approach, a next-hop solicitation message is broadcast by a forwarding node, and all the neighbors receive the message.

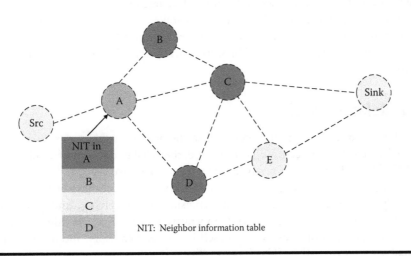

Figure 5.1 Illustration of neighbor information table in a transmitter-oriented approach.

First, the hop count of a neighbor candidate to the sink should be less than that of the forwarding node. Then, each neighbor's eligibility as a next hop is determined according to its remaining energy level, which is in turn reflected into the *temporal gradient*. The next hop candidate with the least temporal gradient (or highest remaining energy) will reply with a next-hop response message with the shortest back-off time. Without central coordination, the candidate with the lowest *temporal gradient* is selected to deliver data packets toward the sink and suppress the other candidates.

The receiver-oriented approach employed by RLRR has the following advantages: (1) the protocol is stateless. Beaconing of a hello message is not needed to update the neighbors' energy information periodically. (2) The energy information is used locally by each next-hop candidate in making its load-balanced routing decision and, hence, is always accurate and up to date.

If there are no next-hop candidates whose hop count to the sink is smaller, RLRR utilizes peer neighbors whose hop count is equal to that of the forwarding node, which can be exploited for reliability and load balancing while guaranteeing loop-free routing. Our receiver-oriented idea is close to Geographic Random Forwarding (GeRaF) [9] and Extreme Opportunistic Routing (ExOR) [10], where efficient methods of using multireceiver diversity for packet forwarding are explored. However, unlike GeRaF and ExOR, RLRR does not rely on geographical information provided by expensive GPS devices.

We carry out extensive simulations to show that RLRR mostly achieves higher reliability than energy-efficient differentiated directed diffusion (EDDD) [5], directed diffusion (DD) [8], and Geographical and Energy-Aware Routing (GEAR) [2]. More importantly, RLRR also exhibits a longer network lifetime. The overall performance gain of RLRR, taking into account reliability, lifetime, and data delivery latency, increases as the link failure rate increases.

The rest of this chapter is organized as follows. We describe RLRR design issues and its algorithm in Sections 5.2 and 5.3, respectively. Simulation model and experimental results are presented in Section 5.4. Finally, Section 5.5 concludes the chapter.

5.1 Related Work

In addition to the background presented in the previous section, our work is also related to cooperative communications and the reliable data transfer scheme in WSNs. We will give a brief review of the work in these two aspects.

A large number of cooperative communication protocols have been proposed recently. Cooperation diversity gains, transmitting, receiving, and processing overheads, are investigated by [11]. Cooperative issues across the different layers of the communication protocol stack, self-interested behaviors, and possible misbehaviors are explored in [12]. Lin and Wong [13] proposed a cooperative relay framework that accommodates the physical, medium access control (MAC), and network layers for wireless ad hoc networks. In the network layer, diversity gains can be achieved by selecting two cooperative relays based on the average link signal-to-noise ratio (SNR) and the two-hop neighborhood information. A cooperative communication scheme combining relay selection with power control is proposed in [14], where the potential relays compute individually the required transmission power to participate in the cooperative communications. A variety of cooperative diversity protocols are proposed by [15]—namely, amplify-and-forward, decode-and-forward, selection relaying, and incremental relaying. The performance of the protocols in terms of outage events and associated outage probabilities are evaluated, respectively. Coded cooperation [16] integrates cooperation with channel coding and works by sending different parts of each user's code word via two independent fading paths. Hunter and Nosratinia [17] and Erkip et al. [18] implemented a cooperation strategy for mobile users in conventional code division multiple access (CDMA) systems, in which users are active and use different spreading code to avoid interferences. In [19], distributed cooperative protocols, including random selection, received SNR selection, and fixed priority selection, are proposed for cooperative partner selection. The outage probability of the protocols are analyzed, respectively. CoopMAC, a cooperative MAC protocol for IEEE 802.11 wireless networks, is presented by [20]. CoopMAC can achieve performance improvements by exploiting both the broadcast nature of the wireless channel and cooperative diversity.

Research efforts studying the issue of reliable data transfer in WSNs [21,22,23,24,25,26] are increasing. In these studies, hop-by-hop recovery are increasing [21,22], end-to-end recovery [26], and multipath forwarding [23,24,25] are the major approaches to achieving the desired reliability. Pump-Slowly, Fetch-Quickly (PSFQ) [21] works by distributing data from source nodes at a relatively slow pace and allowing nodes experiencing data losses to recover any missing segments from immediate neighbors aggressively. PSFQ employs hop-by-hop recovery instead of end-to-end recovery. In [22], the authors proposed Reliable Multi-Segment Transport (RMST), a transport protocol that provides guaranteed delivery for application requirements. RMST is a selective NACK-based protocol that can be configured for in-network caching and repair. In [23], multiple disjoint paths are set up first, and then multiple data copies are delivered using these paths. In [24], a protocol called ReInForM (reliable information forwarding using multiple paths in sensor networks) is proposed to deliver packets at a desired level of reliability by sending multiple copies of each packet along multiple paths from sources to sink. The number of data copies (or the number of paths used) is dynamically determined depending on the probability of channel error. Instead of using disjoint paths, GRAdient Broadcast (GRAB) [25] uses a path interleaving technique to achieve high reliability. It assigns the amount of credit α to each packet at the source. α determines the "width" of the forwarding mesh and should be large enough to ensure robustness but not to cause excessive energy consumption. It is worth noting that, although GRAB [25] also exploits data broadcasting to attain high reliability, it may not be energy-efficient because it may involve many next-hop nodes in order to achieve good reliability and an unnecessarily large number of packets may be broadcast. Considering the asymmetric many-to-one communication pattern from sources to sink in some sensor applications, data packets collected for a single event exhibit high redundancy. Thus, some reliable techniques [21,22] proposed for WSN would either be unnecessary or spend too many resources on guaranteeing 100% reliable delivery of data packets. Exploiting the fact that the redundancy in sensed data collected by closely deployed sensor nodes can mitigate channel errors and node failures, Event-to-Sink Reliable Transport (ESRT) [26] intends to minimize the total energy consumption while guaranteeing the end-to-sink reliability. In ESRT, the sink adaptively achieves the expected event reliability by controlling the reporting frequency of the source nodes. However, in a case where many sources are involved in reporting data simultaneously to ensure some reliability (e.g., in a highly unreliable environment), the large number of communications are likely to cause congestion.

5.2 RLRR Design Issues

5.2.1 Accurate and Up-to-date Energy Information

In local load-balancing protocols, beaconing is required periodically for setting up energy information tables. During the interval between two beacons, the energy information stored in the table does not reflect the actual energy information, since sensor nodes likely consume energy continuously over time. Thus, the interval of broadcasting such a hello message provides a trade-off between control overhead and timeliness of local energy information. In contrast, with the receiver-oriented approach in RLRR, a neighbor node uses its own energy information, which is always accurate and up to date, to evaluate its eligibility to be selected as a next-hop node.

5.2.2 Load Balancing

We assume that every node starts with the same energy level corresponding to full battery capacity. In RLRR, the current energy levels (remaining battery capacities) of the sensor nodes are discretized into integer-valued quantized energy levels (*QELs*). Given the example shown in Figure 5.2, assuming the full energy level (E_{max}) of a battery is equal to 10,000 and the *unit energy* (the unit of the quantization, E_{unit}) is equal to 2000, then the maximum value of *QEL* is ($QEL_{max} = [\frac{E_{max}}{E_{unit}}] = 5$. In this chapter, we do not differentiate the energy levels of sensor nodes with the same *QEL*. For example, both energy levels 6500 and 6750 have the same *QEL* of 4. With effective load balancing in a WSN, the sensor nodes close to one another (e.g., within one hop distance) will have similar *QELs* after an extended period of network operation, because neighbors with higher *QELs* will be selected to forward

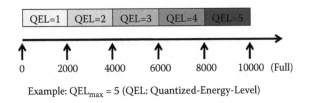

Figure 5.2   Illustration of node energy model used in RLRR.

data until their *QELs* are decreased to levels no higher than those of other neighbors. The larger the range of *QELs*—that is, the smaller the unit energy used in the quantization of the energy levels—the better the load-balancing performance should be. In this case, a longer expected lifetime is likely to be achieved, but at the expense of a higher control overhead to carry out more frequent route oscillations. Thus, QEL_{max} should be optimized to achieve the best trade-off between load balancing and control overhead.

5.2.3 Reliability

With the receiver-oriented approach, the property of broadcasting is exploited to attain high reliability. In RLRR, the source node and any intermediate sensor node broadcast a route selection message. Neighbors that receive the route selection message successfully have the responsibility of choosing the next hop among themselves. If no such available neighbor is found, the node will mark itself a dead-end node and inform the upstream node that it must discover a new route that bypasses the dead end. Especially, RLRR exploits peer neighbors whose hop count is equal to that of the upstream node to increase reliability. However, RLRR faces the challenge of maintaining loop freedom when peer neighbors are exploited.

5.2.4 Loop Freedom

In order to guarantee loop freedom, many routing schemes based on neighbor information only adopt the set of *minimum hop count* nodes as backup next-hop nodes to counteract frequent route failures. A minimum hop count node has a hop count to the sink that is 1 less than the hop count of the current node. Therefore, these schemes exclude the neighbors whose hop counts are the same as that of the current node (i.e., peer neighbors) as potential next-hop nodes. In RLRR, the number of backup nodes is exploited to the maximum extent possible by also involving peer neighbors to route data packets in order to achieve better load balancing and reliability. With the receiver-oriented approach, loop freedom is guaranteed with no additional control overhead, as will be explained in detail in Section 5.3.4.

5.2.5 Low-Cost Sensor Design

Traditional sensor routing protocols usually require a sensor node to maintain the information of multiple neighbors (e.g., backup routes and energy levels). In very large-scale and dense WSNs, the amount of such information may pose an additional challenge for the sensor nodes with low storage capacity. However, with RLRR, sensor nodes do not need to store any additional routing and energy-related information except for the identifier of its next-hop node and its upstream node for each flow. Though stateless geographical routing schemes also do not need to set up route tables, they need to obtain geographical information using GPS devices. By comparison, RLRR does not need any geographical information to achieve stateless routing.

5.3 The RLRR Protocol

5.3.1 The RLRR Mechanism

In RLRR, each node has a *flow entry*, which indicates the identifier of its next-hop node for forwarding data to the sink. Initially, a sink floods interest packets to the network. Each sensor sets up its hop count gradient to the sink. Sensor(s) matching the interest will become the source node(s) [8]. Unlike minimum hop count-based routing schemes, the flow entry is not set up during interest flooding in RLRR, since load balancing cannot be attained simply by considering hop count. Instead, the flow entries of all the sensor nodes are still empty after interest flooding.

We denote a forwarding node (the source or an intermediate node) by h. The arrival of a sensory data packet (from the application layer of the source node or from the upstream node) triggers h to check its flow entry. Since the flow entry does not exist initially, h stores the data, starts a route selection process immediately to set up the flow entry, and then transmits the stored data to the selected next-hop node. As illustrated in Figure 5.3(a), suppose node i with a QEL of 5 is selected as the next-hop node of node h. After the flow entry is set up, data packets will be unicast directly to the next-hop node recorded in the flow entry.

As time goes on, node i will consume its energy faster than its neighbors. To achieve load balancing, node i should keep track of its own QEL in order to prevent excessive energy consumption for packet forwarding. When its QEL is decreased by 1, node i asks its upstream node h to select a new next-hop node. In the example shown in Figure 5.3(b), when the QEL of node i changes from 5 to 4, it unicasts a next-hop-reselection message (RESEL) to its upstream node h. Upon receiving the RESEL, h deletes its current flow entry and initiates route reselection. Assuming node j is selected due to its higher energy level, node i will be replaced by node j as the new next-hop node of h.

In addition to balancing the energy consumption, route reselection is also triggered to recover a link failure. In the example shown in Figure 5.3(c), node h fails to deliver a data packet to node i according to the existing flow entry and receives

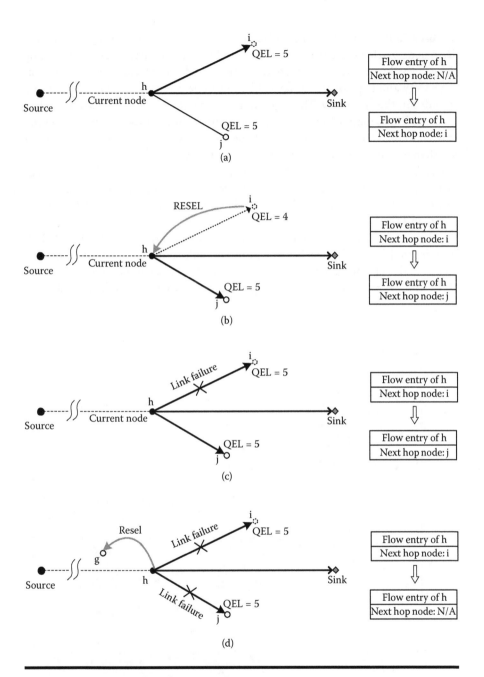

Figure 5.3   Setup/update flow entry in RLRR.

feedback information from its MAC layer that indicates a transmission failure. Then node *h* deletes its current flow entry and initiates route reselection. Assuming the wireless link to node *j* is in good condition and other factors (such as remaining energy and hop count) are favorable, this node is selected as the next hop and recorded in the flow entry. The original next-hop node *i* is now replaced by node *j*.

In addition, route selection/reselection (denoted by Sel/Resel, respectively) itself may fail. For example, if all the eligible neighbors (whose hop count to the sink is less than or equal to that of *h*) of node *h* have either depleted their energies or failed, node *h* becomes a dead-end node. In this case, node *h* transmits a RESEL message to its upstream node (e.g., node *g* in Figure 5.3(d)), which triggers a new route Reselection by node *g*, and so forth. The flowchart of the basic RLRR Protocol is shown in Figure 5.4.

Figure 5.5 shows the mechanism of receiver-oriented route Sel/Resel in RLRR. In order to deliver a data packet to a next-hop node in dynamic network environments, node *A* broadcasts a probe message at first. The neighbor nodes (i.e., nodes *B*, *C*, and *D*) that receive this message and are closer to the sink than node *A*,

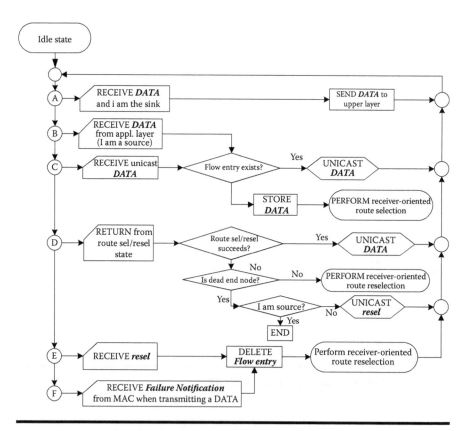

Figure 5.4　Flowchart of the basic RLRR protocol.

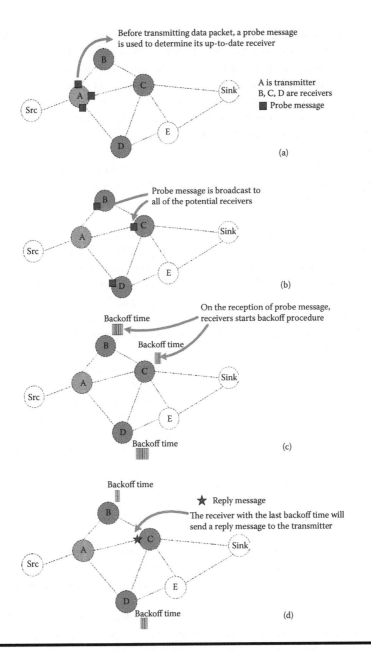

Figure 5.5   Illustration of receiver-oriented mechanism in RLRR.

will start their back-off timers, as shown in Figure 5.5(c). They are also called *live candidates* (LCs), such that the links between the transmitter and the candidates are in good status. Since multiple LCs usually start their back-off timers (denoted by TG-Timers) simultaneously, the one (i.e., node C, as shown in Figure 5.5(d)) with the lowest TG (*time gradient*) will expire first and becomes a *reserved next hop* (RNH), which means it is highly likely to be selected as the next-hop node later. Note that the TG of an individual LC indicates its eligibility level to be selected as the next hop.

Ideally, all the LCs except the RNH should cancel their TG-Timers and delete the packet from their forwarding buffers when the RNH's TG-Timer expires. To achieve this, RLRR operates as follows:

1. RNH broadcasts a reply message (REP) to node h.
2. If node h receives the REP from RNH, it will broadcast a selection message (SEL) with the identifier of the RNH, and start the selection-retransmission-timer (SEL-ReTx-Timer). To guarantee that only one LC is selected as the next-hop node, node h only accepts the first REP sent by the RNH while ignoring the later ones. Note that the LCs overhearing the REP will back out (i.e., cancel their TG-Timers and drop the data from their caches) instantly.
3. If the RNH receives the SEL, it becomes the next-hop node and relays the data by broadcasting. When other LCs receive the SEL, they will cancel their TG-Timers and drop the data sent by their forwarding caches.
4. If node h receives the broadcast data from its next-hop node (the above RNH), it will cancel its SEL-ReTx-Timer. Otherwise, it will rebroadcast the SEL when the SEL-ReTx-Timer expires and will start the timer again until the retry limit is reached.
5. If the RNH receives a retransmitted SEL, it will unicast a selection-reply message (SEL_REP) to node h.
6. If node h receives SEL_REP, it will cancel its SEL-ReTx-Timer.

Note that in Step 1, two (or more) LCs with similar TGs broadcast their REPs simultaneously. If collision happens, neither of the LCs will be selected, and other LCs that broadcast REPs later, when their TG-Timers expire, will be selected.

Furthermore, in the above Step 2 it is possible that the SEL may collide with a new REP from another LC, which would cause the following two disadvantages: (a) RNH may fail to receive the SEL; (b) other LCs (non-RNH nodes) do not delete the data from their caches quickly enough. Case (b) only increases data caching time and control overhead, while case (a) will cause the failure of the current data delivery if left without any measure. To ensure that the RNH receives the SEL at least once, node h should send the SEL again when SEL-ReTx-Timer expires.

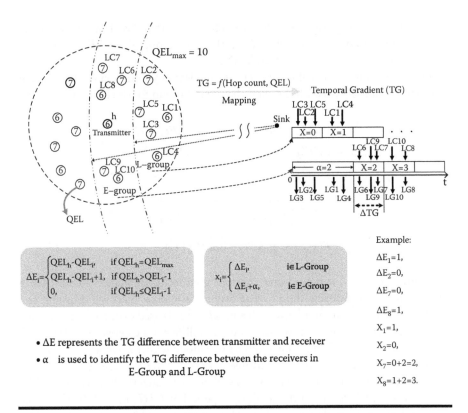

Figure 5.6 Converting energy and hop count information into temporal gradient.

5.3.2 Time Gradient Calculation

In Figure 5.6, the *LCs* of node *h* are divided into two groups: (1) the less-hop-count group (*L-Group*), consisting of *LCs* that are one hop closer to the sink than node *h*; and (2) the equal-hop-count group (*E-Group*), consisting of the *LCs* having the same hop count as node *h*. Obviously, the *LCs* in *L-Group* should have a higher priority than those in the *E-Group*. In Figure 5.6, the *L-Group* includes nodes *LC*1, *LC*2, *LC*3, *LC*4, and *LC*5; and the *E-Group* includes nodes *LC*6, *LC*7, *LC*8, *LC*9, and *LC*10.

Recall that node *h* broadcasts a PROB to initiate route Sel/Resel. The PROB contains the *QEL* and hop count of node *h*. In Figure 5.6, QEL_{max} is equal to 10, and the *QEL* of each *LC* is indicated by the number in the respective circle. Upon receiving the PROB, an *LC* first decides to which group it belongs. Then it calculates the gap between its own *QEL* and the upstream node *h*'s *QEL*, which is denoted by ΔE. Since time gradient (i.e., *TG*) determines the delay of sending a REP back to *h*, its value has a large impact on the data latency. In order to make *TG* as small as possible while achieving sufficient differentiation among all the LCs, we should avoid using large *TG* values to differentiate the LCs. Thus, we adopt ΔE instead of *QEL* to differentiate the *LCs* in the same group, since ΔE can be much smaller than *QEL* in a load-balanced WSN.

Let TG_i denote the *TG* of node *i*. TG_i is calculated by Equation (5.1), where *x* is a parameter that reflects both ΔEs and the type of group to which an *LC* belongs.

$$\Delta E_i = \begin{cases} QEL_h - QEL_i, & \textit{if } QEL_h = QEL_{\max} \\ QEL_h - QEL_i + 1, & \textit{if } QEL_h > QEL_i - 1 \\ 0, & \textit{if } QEL_h \le QEL_i - 1 \end{cases}$$

$$x_i = \begin{cases} \Delta E_i, & i \in L - Group \\ \Delta E_i + \alpha & i \in E - Group \end{cases}$$

$$TG_i = f(x_i) = x_i \times \Delta TG + rand(\Delta TG)$$

(5.1)

In Equation (5.1), α is a positive constant used to differentiate between LCs in different groups by favoring the *L-Group* over the *E-Group*, ΔTG is a constant, and $rand(\Delta TG)$ is a random value between 0 and ΔTG that is used to differentiate the *LCs* that have the same *x*. In other words, it is used to differentiate between multiple *LCs* that have the same *QEL* and belong to the same group (either *L-Group* or *E-Group*).

ΔTG should be set as small as possible to decrease the Sel/Resel delay, but if it is set too low, collisions of REP messages will occur frequently, because many LCs will try to send REPs within the small time period of ΔTG. Thus, ΔTG should be set according to the node density. Let *N* be the total number of sensor nodes in a WSN that has an area *A*. The node density of the WSN is equal to $\delta = \frac{N}{A}$. Let *r* be the transmission range of a sensor node. Roughly, *LCs* are located within approximately one third of the whole transmission range in Figure 5.6. Then the number of *LCs* can be approximated by:

$$L = \frac{1}{3} \cdot \pi \cdot r^2 \cdot \delta$$

(5.2)

Among the *LCs* in the same group, on the average, half of them will have the same *QEL* in a *load-balanced* WSN. Let *S-Group* denote the set of the *LCs* with the same *QEL* in the same group (i.e., *L-Group* or *E-Group*). Our goal is to make a contention time long enough to differentiate the *LCs* in the same *S-Group*. Let T_{REP} be the average time to successfully deliver a REP message. In order to minimize collisions with other *LCs* in the same *S-Group*, at least T_{REP} should be reserved for each *LC*. Thus, ΔTG is approximately equal to:

$$\Delta TG = \frac{L}{2} \cdot T_{REP}.$$

(5.3)

Here, $\frac{1}{2}$ is the average number of *LC*s in an *S-Group*. Let α be 2. In the example shown in Figure 5.6, we can get four *S-Group*s: *LC*2, *LC*3, and *LC*5 with the $x = 0$; *LC*1 and *LC*4 with $x = 1$; *LC*6, *LC*7, and *LC*9 with $x = 2$; *LC*8 and *LC*10 with $x = 3$. The *TG*s of the *LC*s in each *S-Group* are randomly distributed over a range of ΔTG. In the example in Figure 5.6, the increasing order of the *TG*s of all the *LC*s is: *TG*3, *TG*2, *TG*5, *TG*1, *TG*4, *TG*6, *TG*9, *TG*7, *TG*10, and *TG*8. It corresponds to the decreasing order of the *LC*'s eligibility level as *h*'s next-hop node: *LC*3, *LC*2, *LC*5, *LC*1, *LC*4, *LC*6, *LC*9, *LC*7, *LC*10, and *LC*8. As time goes on, the *TG-Timer* of *LC*3 will expire first, which causes *LC*3 to be selected as the next-hop node.

5.3.3 Solving the Dead-End Problem

The so-called dead-end problem [28] arises when a packet is forwarded to a local optimum (that is, a node with no neighbor that has a closer hop distance to the destination). The problem can be solved as follows: (1) If node *h* does not receive any REP until its *NoREP-Timer* expires, it will mark itself as an unavailable node and unicast a RESEL to its upstream node *u*. An unavailable node will not participate in route Sel/Resel until the sink floods a new control message. The frequency of the sink flooding a control message should be traded off between control overhead and the timeliness of mitigating the dead-end problem; (2) On receiving the RESEL, node *u* initiates route reselection and finds a new next hop to replace node *h*.

5.3.4 Loop Freedom in RLRR

Exploiting multiple backup nodes or multipath for data delivery can increase reliability. In general, the nodes in the equal-hop-count group (*E-Group*) are not used as backup nodes to guarantee loop freedom in conventional routing schemes in ad hoc and sensor networks. By comparison, in RLRR, the *LC*s in the *E-Group* are exploited to achieve better performance in terms of both load balancing and reliability. While using the *LC*s in the *E-Group*, it is critical to ensure that an *LC* in the *E-Group* not be selected as a next hop again by another *LC* in the same *E-Group*. The receiver-oriented approach of RLRR makes this goal easily achievable with no additional control overhead, as illustrated in Figure 5.7.

In Figure 5.7(a), assume that node *a* is the only minimum-hop-count neighbor of node *h*, and it fails. Then, node *h* initiates route Resel by broadcasting a PROB message and starting a *Drop-PROB-Timer*. The route Resel results in node *h* selecting its peer neighbor node *b* (in the *E-Group*) as the next-hop node. Assuming that the flow entry of node *b* does not exist, in Figure 5.7(b), node *b* initiates route selection. Note that both *h* and *c* are peer neighbors of node *b*. When node *b* broadcasts a PROB and node *h* replies first, node *h* is selected as *b*'s next-hop node and a loop is formed.

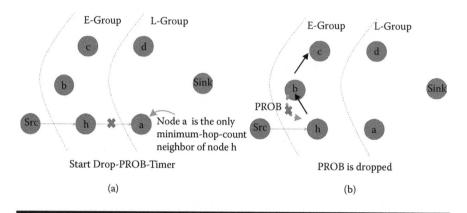

Figure 5.7   Illustration of guaranteeing loop freedom.

To prevent node *h* from being selected as a next-hop node by its peer neighbors, it ignores the PROBs and never participates in the selection until its *Drop-PROB-Timer* expires. When the *Drop-PROB-Timer* of node *h* expires, it can participate in the next-hop selection process again. In general, the time value of *Drop-PROB-Timer* ($T_{\text{Drop-PROB-Timer}}$) should be long enough, such as:

$$T_{Drop-PROB-Timer} = N_{E-Group} \cdot t_{sel}. \tag{5.4}$$

In Equation (5.4), $N_{E-Group}$ denotes the maximum number of *LC*s in the *E-Group*. t_{sel} denotes the time for one-hop route selection. Considering the worst case, where all the *LC*s in the *E-Group* have no minimum-hop-count neighbors, each of them will initiate route selection once and find a peer neighbor in the same *E-Group* as its next-hop node. Then the accumulated time for all of the route selection attempts is equal to Equation (5.4), which guarantees that a node (e.g., node *h*) initiating route Sel/Resel will never be selected as a next-hop node of any other *LC*s in the same *E-Group*. Thus, loop freedom is guaranteed.

5.4 Performance Metrics

In order to demonstrate the performance of RLRR, we compare it with several representative existing routing protocols for WSNs by extensive simulation studies.

We chose a global load-balancing scheme (i.e., EDDD [5]), a local load-balancing scheme (i.e., GEAR [2]), and a non-load-balancing scheme (i.e., DD [8]) to compare with RLRR. We implement the HGR protocol and perform simulations using OPNET Modeler [32,33]. The sensor nodes are battery operated except for the sink, which is assumed to have an infinite energy supply. The network with 800 nodes is uniformly deployed over a 500 m × 500 m field. As in [30], we let one sink stay at a corner of the field and one source node be located at the diagonal corner. Each source node generates sensed data packets at a constant bit rate with a 5-second interval between packets (1K Bytes each). As in [6], we use IEEE 802.11 DCF as the underlying MAC, and the radio transmission range (R) is set to 45 m. The data rate of the wireless channel is 2 Mbps. All messages are 128 bytes in length. We assume both the sink and sensor nodes are stationary. In DD, EDDD, and RLRR, the sink will initiate interest flooding to carry out a new task. Interest packets are propagated on a hop-by-hop basis throughout the network. Among the target sensor nodes, while several nodes may match the interest, only one of these nodes will become a source node for each instance of interest flooding. We assume that a mechanism exists to elect one source node among several nodes that matches the interest—for example, based on the remaining energy. In addition to the initial interest flooding, the sink also floods the interest packet periodically to update stale information in terms of hop count and energy. Since RLRR does not rely on periodic flooding for local repair, the sink only floods interest once until the network lifetime is reached. We employ the energy model used in [30,31] and the link failure model used in [29]. For each set of results, we simulate the WSN 60 times with the specific set of parameters and different random seeds.

In this section, five performance metrics are defined:

- *Number of Successful Data Deliveries during Lifetime*—This is the number of data packets delivered to the sink before the network lifetime is reached. It is denoted by n_{data}, which is also used as an indication of the lifetime in this chapter.
- *Packet Delivery Ratio*—This is the ratio of the number of data packets delivered to the sink to the number of packets generated by the source nodes.
- *Average End-to-end Packet Delay*—This includes all possible delays during data dissemination, caused by queuing, retransmissions due to collisions at the MAC layer, and transmission time.
- *Energy Consumption per Successful Data Delivery*—This is denoted by e. It is the ratio of network energy consumption to the number of data packets delivered to the sink during the network lifetime. The network energy consumption includes all the energy consumption due to transmitting and receiving during the simulation. As in [29,30,6,31], we do not account for energy consumption during the idle state, since this element is approximately the same for all the schemes considered.
- *Number of Control Messages per Successful Data Delivery*—This is denoted by n_{ctrl} and is the ratio of the number of control messages transmitted to the number of data packets delivered to the sink during the network lifetime.

We use n_{data} as an approximate indication of the network lifetime. If the packet delivery ratio is 100%, then n_{data} is exactly proportional to the network lifetime, due to the CBR traffic model used in the simulations. We believe that n_{data} is the most important metric for WSNs.

5.4.1 Effects of ΔTG

In these experiments, we change ΔTG from 2 ms to 20 ms by the step size of 2 ms. In Figure 5.8(a), n_{data} increases as ΔTG is increased, since the larger ΔTG is, the less collisions will happen, and the more data packets will be delivered successfully to the sink.

In Figure 5.8(b), when ΔTG is small, end-to-end data delay of RLRR is high. The smaller ΔTG is, the more likely will the REPs transmitted by LCs with similar TGs collide, which causes LCs with lower TGs not to win the opportunity to become a next-hop node. With increasing ΔTG, the delay decreases and reaches its minimum value when ΔTG is equal to 8 milliseconds. It is unnecessary to increase ΔTG more if the value is large enough to differentiate the LCs in the same S-$Group$, since a large ΔTG also increases the time for route Sel/Resel. Thus, when ΔTG goes beyond 8 ms, the delay begins to increase again.

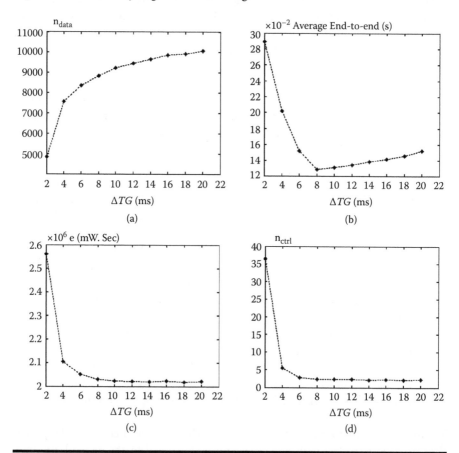

Figure 5.8 The impact of ΔTG on: (a) n_{data}; (b) end-to-end delay; (c) e, and (d) n_{ctrl}.

In Figures 5.8(c) and 5.8(d), both e and n_{ctrl} decrease with ΔTG increasing. The larger ΔTG is, the less likely it is that a collision will happen, and thus the control overhead decreases.

5.4.2 Comparison of RLRR, EDDD, DD and GEAR with Variable Link Failure Rates

In this section, we change the link failure rate from 0 to 0.5 by the step size of 0.05. Figure 5.9(a) shows that the packet delivery ratios of EDDD and DD are more sensitive to link failures than those of GEAR and RLRR, and EDDD has the lowest reliability because the load-balanced path is not robust to link failures, since the failure of any link along the path will cause data delivery failure. RLRR yields higher reliability than GEAR because it exploits the *E-Group* for alternating routing. In most cases, the numbers of nodes in the *L-Group* and *E-Group* are larger than the number of backup nodes in GEAR. Thus, RLRR keeps achieving more than 90% packet delivery ratio until the link failure rate is larger than 0.35.

In Figure 5.9(b), when the link failure rate is 0, n_{data} of EDDD is larger than that of RLRR and GEAR, which illustrates the advantage of global load balancing (EDDD) over local load balancing (RLRR, GEAR) in reliable environments. Note that n_{data} is closely related to the network lifetime. Since DD has no load-balancing mechanism, its lifetime is the lowest. With an increasing link failure rate, RLRR exhibits consistently higher reliability and n_{data} than the other schemes, which shows that the proposed receiver-oriented scheme can achieve load balancing with a lower control overhead and handle link failure better than conventional transmitter-oriented schemes.

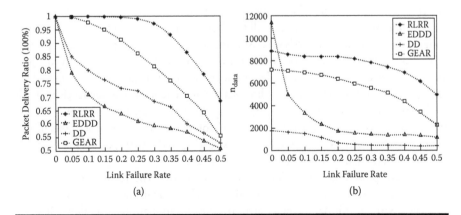

Figure 5.9 The impact of link failure rate on: (a) reliability and (b) lifetime (n_{data}).

According to the simulation results, we observe the following: (1) lifetime is greatly prolonged if a load-balancing mechanism is adopted (e.g., EDDD, RLRR, and GEAR vs. DD); (2) in a reliable environment, a global load-balancing scheme exhibits a longer lifetime than local load-balancing schemes (e.g., EDDD vs. RLRR and GEAR); and (3) RLRR exhibits more consistent and relatively higher reliability and longer lifetime than EDDD, DD, and GEAR in unreliable environments.

5.5 Conclusion

Routing protocols in wireless sensor networks (WSNs) typically employ a transmitter-oriented approach in which the next-hop node is selected based on neighbor or network information. This approach incurs a large overhead when the accurate neighbor information is needed for efficient and reliable routing. Additionally, in unreliable communication environments, traditional routing protocols may fail to deliver data in a timely manner since global route discovery may be needed to handle link failures.

In this chapter, a novel opportunistic routing protocol (denoted by RLRR) is proposed for delivering data in a load-balancing and reliable fashion. In the proposed scheme, an intermediate node solicits next-hop candidates, each of which is to respond with its own back-off time, dubbed a temporal gradient (TG). In RLRR, the energy and hop count information of each live candidate (LC) is converted to a TG that is used to evaluate the eligibility of the node as a next-hop node. The set of LCs of a forwarding node h includes all neighbors of transmitter, whose hop counts to the sink are less than or equal to that of transmitter. To perform route selection, the transmitter broadcasts a probe message (PROB) that is received by its LCs. Each LC sets its TG-$Timer$ to the calculated TG value and sends a reply message (REP) back to the transmitter when its TG-$Timer$ expires. The LC that originated the first reply message received by the transmitter is selected as the next-hop node. The best LC will have the least TG; therefore, its TG-$Timer$ will expire first among all the LCs and it will be selected as the next-hop node. In this way, the next hop is selected without any central coordination on a packet-by-packet basis. Thus, each node does not need to maintain any neighbor information. The remaining energy level used to determine the TG is always accurate and up to date. The upstream node of a broken link broadcasts a route request message received by all the live neighbors with a good link. By taking this local approach, route repair is fast and reliability is enhanced even in highly unreliable environments. Furthermore, neighbor nodes whose hop count is less than the soliciting node participate in the next-hop selection process with loop-free operation guarantee.

We have presented simulation results to show that the related parameters of the protocol need to be selected carefully to achieve load balancing with energy efficiency while minimizing the control overhead. Simulations also show that the proposed protocol achieves relatively longer network lifetime and higher reliability than other existing schemes.

References

[1] J. Al-Karaki and E. Kamal. Routing techniques in wireless sensor networks: A survey. *IEEE Personal Communications*, 11(6):6–28, December 2004.

[2] Y. Yu, R. Govindan, and D. Estrin. *Geographical and energy-aware routing.* UCLA Computer Science Department Technical Report UCLA/CSD-TR-01-0023, May 2001.

[3] H. Dai and R. Han. A node-centric load balancing algorithm for WSNs. In *Proceedings of IEEE GLOBECOM*, Vol. 1, pp. 548–552, December 2003.

[4] J. Gao and L. Zhang. Load balancing shortest path routing in wireless networks. In *Proceedings of IEEE INFOCOM*, Vol. 2, pp. 1098–1107, March 2004.

[5] M. Chen, T. Kwon, and Y. Choi. Energy-efficient differentiated directed diffusion for real-time traffic in wireless sensor networks. *Elsevier Computer Communications*, 29(2):231–245, January 2006.

[6] M. Chen, V. Leung, S. Mao, and Y. Yuan. Directional geographical routing for real-time video communications in wireless sensor networks. *Elsevier Computer Communications*, 30(17):3368–3383, November 2007.

[7] M. Chen, V. Leung, S. Mao, and T. Kwon. Receiver-oriented load-balancing and reliable routing in wireless sensor networks. *Wireless Communications and Mobile Computing Journal*, 9(3):405–416, March 2009.

[8] C. Intanagonwiwat, R. Govindan, and D. Estrin. Directed diffusion: A scalable and robust communication paradigm for sensor networks. In *Proceedings of ACM MobiCom 2000*, pp. 56–67, Boston, MA, August 2000.

[9] M. Zorzi and R. Rao. Geographic Random Forwarding (GeRaf) for ad hoc and sensor networks: Multihop performance. *IEEE Transactions on Mobile Computing*, 2(4):337–348, 2003.

[10] S. Biswas and R. Morris. ExOR: Opportunistic multi-hop routing for wireless networks. In *Proceedings of ACM SIGCOMM*, pp. 133–144, 2005.

[11] Y. W. A. K. Sadek and K. R. Liu. When does cooperation have better performance in sensor networks? In *Proceedings of the 3rd IEEE Sensor and Ad Hoc Communications and Networks (SECON'06)*, pp. 188–197, 2006.

[12] E. G. M. Conti and G. Maselli. Cooperation issues in mobile ad hoc networks. In *Proceedings of the 24th International Conference on Distributed Computing Systems Workshops(ICDCSW'04)*, pp. 803–808, 2004.

[13] J. S. Y. Lin and V. W. Wong. Cooperative protocols design for wireless ad-hoc networks with multi-hop routing. *Mobile Networks and Applications*, 4(2):143–153, 2009.

[14] J. C. Z. Zhou, S. Zhou, and S. Cui. Energy-efficient cooperative communication based on power control and selective single-relay in wireless sensor networks. *IEEE Transactions on Wireless Communications*, 7(8):3066–3078, 2008.

[15] D. N. T. J. N. Laneman and G. Wornell. Cooperative diversity in wireless networks: Efficient protocols and outage behavior. *IEEE Transactions on Information Theory*, 50(12): 3062–3080, 2004.

[16] T. Hunter and A. Nosratinia. Cooperation diversity through coding. In *Proceedings of the IEEE International Symposium on Information Theory (ISIT'02)*, pp. 220–225, 2002.

[17] E. Erkip, A. Sendonaris, and B. Aazhang. User cooperation diversity-Part I: System description. *IEEE Transactions on Communications*, 51(11):1927–1938, 2003.

[18] E. Erkip, A. Sendonaris, and B. Aazhang. User cooperation diversity-Part II: Implementation aspects and performance analysis. *IEEE Transactions on Communications*, 50(11):1939–1948, 2003.

[19] T. Hunter and A. Nosratinia. Distributed protocols for user cooperation in multi-user wireless networks. In *Proceedings of the 47th IEEE Annual Global Telecommunications Conference (GLOBECOM'04)*, pp. 3788–3792, 2004.

[20] Z. L. E. E. P. Liu, Z. Tao, and S. Panwar. Cooperative wireless communications: A cross-layer approach. *IEEE Wireless Communications*, 13(4): 84–92, 2006.

[21] C. Wan, A. Campbell, and L. Krishnamurthy. Pump-Slowly, Fetch-Quickly (PSFQ): A reliable transport protocol for sensor networks. *IEEE Journal of Selected Areas in Communications*, 23:862–872, April 2005.

[22] F. Stann and J. Heidemann. RMST: Reliable data transport in sensor networks. In *Proceedings of the IEEE International Workshop on Sensor Network Protocols and Applications*, pp. 102–112, May 2003.

[23] D. Ganesan, R. Govindan, S. Shenker, and D. Estrin. Highly resilient, energy efficient multipath routing in wireless sensor networks. *Mobile Computing and Communications Review (MC2R)*, 1(2):10–24, 2002.

[24] B. Deb, S. Bhatnagar, and B. Nath. ReInForM: Reliable information forwarding using multiple paths in sensor networks. *IEEE LCN*, 406–415, October 2003.

[25] F. Ye, G. Zhong, S. Lu, and L. Zhang. GRAdient Broadcast: A robust data delivery protocol for large scale sensor networks. *ACM Wireless Networks*, 11(3):285–298, 2005.

[26] Y. Sankarasubramaniam, O. Akan, and I. Akyildiz. ESRT: Event-to-Sink Reliable Transport in wireless sensor networks. In *Proceedings of ACM MobiHoc'03*, pp. 177–188, June 2003.

[27] I. Stojmenovic. Position-based routing in ad hoc networks. *IEEE Communications Magazine*, 40(7):128–134, July 2002.

[28] L. Zou, M. Lu, and Z. Xiong. A distributed algorithm for the dead end problem of location-based routing in sensor networks. *IEEE Trans. Vehicular Technology*, 54:1509–1522, July 2005.

[29] M. Chen, T. Kwon, S. Mao, Y. Yuan, and V. Leung. Reliable and energy-efficient routing protocol in dense wireless sensor networks. *International Journal on Sensor Networks*, 4(12):104–117, August 2008.

[30] M. Chen, T. Kwon, Y. Yuan, Y. Choi, and V. C. M. Leung. Mobile agent-based directed diffusion in wireless sensor networks. *EURASIP Journal on Advances in Signal Processing*, 219–241, 2007.

[31] M. Chen, T. Kwon, S. Mao, and V. Leung. STEER: Spatial-Temporal Relation-Based Energy-Efficient Reliable Routing Protocol in wireless sensor networks. *International Journal on Sensor Networks*, (2):129–141, 2008.

[32] OPNET Technologies, Inc., http://www.opnet.com.

[33] M. Chen. *OPNET network simulation*. Press of Tsinghua University: Beijing, 2004.

Chapter 6

Trace-Based Analysis of Mobile User Behaviors for Opportunistic Networks

---

Wei-Jen Hsu and Ahmed Helmy

Contents

The gaining popularity of portable computing and communication devices gives rise to the proposal of opportunistic communication networks among these wireless-capable devices. There have been several recent studies in the area of opportunistic networking analysis. Some of the work relies on simplistic user models, while other studies are based on relatively small-scale experiments. Both approaches are useful in maintaining simplicity, yet often at the expense of realism.

In this chapter, we first present a framework named *TRACE*, outlining procedural steps to understand user behaviors based on realistic data. This framework provides a procedural approach to analyze large-scale wireless network user data sets collected from operational networks in complex environments (e.g., university and corporate campuses). We follow these steps to understand the individual user mobility process, encounter user dynamics, and highlight the lessons we learned from this approach. We focus on identifying (1) the similarities and differences from the different environments presented by the traces, and (2) the potential applications of the findings on user modeling and protocol design. The aim of the data-driven approach is to develop a fundamental understanding of realistic user behavior in wireless networks. These findings show sharp contrast between the characteristics obtained from empirical data sets and those exhibited by many existing, commonly used, random mobility models with homogeneous nodal behavior.

We first give a short introduction to the importance of mobility analysis to the opportunistic networks in Section 6.1 and introduce our TRACE framework in section 6.2. The data sets we use for the analysis are briefly discussed in Section 6.3. We take two perspectives with different scopes of analysis in an effort to understand user behaviors from the traces. In the first part of the chapter, in Section 6.4, the focus is on individual users and their mobility characteristics. We then widen the scope of analysis and switch the focus to the interaction between users through the analysis of internode encounter patterns in Section 6.5. This approach also displays the richness of information one can obtain from these empirical user traces. Finally, we point out some future research directions in Section 6.6 and offer our conclusions in Section 6.7.

6.1 Background on Opportunistic Networks

In recent years, a keen interest in infrastructure-independent communication has led to the establishment of the research area generally known as *Mobile ad hoc networks* (MANETs) [3]. By definition, MANETs are self-organized, infrastructureless networks, and are considered as stand-alone networks in which the participants exchange information among themselves. Typically, MANETs consist of autonomous devices, and each device plays the role of both an end-host and a router at the same time. While the communication range of individual nodes is limited to its close vicinity due to the nature of the wireless medium, the end-to-end connectivity in the network is provided by the cooperation of its participants, through multihop forwarding, which sometimes involves temporary storage of the messages in the memory of intermediate nodes.

A special class of MANETs, sometimes referred to as *opportunistic networks* or *delay-tolerant networks* (DTNs), is considered to be a challenging scenario for mobile ad hoc networks. In opportunistic networks, frequent, long-lasting disconnections between mobile nodes are considered part of the normal operation of the network (i.e., not an exception), and the nodes must leverage sporadic, sometimes unpredictable *encounter* events between nodes (i.e., when two nodes move within wireless communication range) to exchange control and data messages, enabling networkwide communication. Thus, the performance of opportunistic networks and their protocols are fundamentally determined by the mobility process of the nodes and the encounter events created through mobility. Analysis of the mobility process, therefore, becomes an essential task in opportunistic networks research.

While mobility modeling has been a focus of research and an array of mobility models have been proposed (e.g., see [13] for a summary), few proposals of random mobility models (e.g., random waypoint, random direction, or the Manhattan model [15]) are based on empirical mobile network user data. We think it is imperative to observe *realistic user behaviors* from *operational wireless networks* and understand the mobility process and the encounter patterns between the nodes empirically. This approach has the potential of not only leading to sound mobility and encounter models, which pave the way for performance analysis in opportunistic networks, but also building the necessary understanding to develop *behavior-aware* protocols and services. The need for systematic steps in this line of research led to the TRACE framework we propose in the next section.

6.2 The TRACE Framework

We use the TRACE framework (see Figure 6.1) as the guiding methodology for our realistic analysis of user behaviors. The study starts from an extensive library of *Traces* we first introduce in Section 6.3. We then construct different

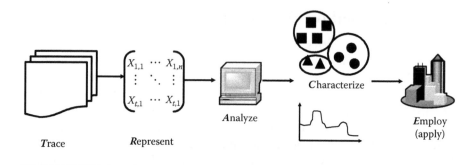

Figure 6.1   Illustration of the TRACE framework.

Representations (distribution curves and graphs, etc.) to *Analyze* various aspects of user behaviors, including individual and encounter dynamics, in Sections 6.4 and 6.5, respectively. There are several common and distinct *Characteristics* of the individual user mobility process and encounter dynamics of users from all environments, including (1) low "online" or usage time of the users, (2) a very limited number of location visits and a skewed visiting time distribution, and (3) a repetitive daily/weekly location visiting pattern; and a Small World pattern emerges from the encounter-relationship graphs. Finally, we will describe the direct inferences of these findings on user modeling and DTN routing performances, and discuss other potential future research directions in Section 6.6. These points display the *Employment* (application) part of the *TRACE* framework.

The TRACE framework serves as the guiding principle for assessing an extensive collection of realistic data to understand user behavioral patterns and to employ this understanding for various tasks (e.g., user modeling or protocol performance analysis). In this chapter we advocate this empirical approach, as it provides realistic understanding and important insights to complement theoretical analyses and small-scale experiments.

6.3 Data Sets Used

In this work we mainly focus on wireless local area network (WLAN) traces collected from university campuses and corporate networks. We obtain wireless traces from various sources, including more than 60,000 distinct users and over 2,000 access points (APs). To our best knowledge, this is the most extensive study of user behavior in wireless networks so far. Many of the traces we use are available at the trace archives established for the purpose of resource sharing in the community (e.g., [1,2]). We summarize some important facts about these traces in Table 6.1

Table 6.1 Statistics of Studied Traces

Trace Source	Unique Users	Unique Aps	Trace Duration	User Type	Analyzed Part in this Work	Users in Analyzed Part	Labels Used in Graphs
MIT[a] [22]	1,366	173	July 20, 2002 to August 17, 2002	WLAN Generic	Whole trace	1,366	MIT
Dartmouth [21]	10,296	623	April 2001 to June 2004	WLAN Generic	July 2003	2,518	Dart-03
				WLAN Generic	April 2004	5,582	Dart-04
				PDA only	April 2004	25	Dart-PDA
				VoIP device	April 2004	63	Dart-VoIP
UCSD[b] [23]	275	518	September 22, 2002 to December 8, 2002	PDA only	September 22, 2002 to October 21, 2002	275	UCSD
USC[c] [20]	4,548	79 ports	April 20, 2005 to Now	WLAN Generic	April 20, 2005 to May 19, 2005	4,528	USC
UF[d]	44,751	728	August 2007 to Now	WLAN Generic	January 14, 2008 to February 13, 2008	32,695	UF

[a] MIT = Massachusetts Institute of Technology
[b] UCSD = University of California, San Diego
[c] USC = University of Southern California
[d] UF = University of Florida

and explain the major issues below. In these traces, each unique mobile node (MN) is identified by its MAC address.*

These traces are chosen to represent different environments and different time periods of the studied WLAN deployments. We study the differences and similarities of user behavior in these traces and try to attribute them to the underlying differences in the corresponding environments as appropriate. The MIT, Dartmouth, University of Southern California (USC), and University of Florida (UF) traces collect measurements of generic WLAN users in a corporate network (the MIT trace) or university campus networks (the other three traces), including all WLAN-capable devices. The UCSD trace is from a specific project to study the behaviors of personal digital assistant (PDA) users. To further compare the association behaviors of smaller, handheld devices (e.g., PDA, Voice over Internet Protocol [VoIP] devices) with generic wireless devices in the same environment, we also separate the PDA and VoIP device users from the generic Dartmouth trace during April 2004, and study their behavior specifically.†

In these traces, the current location of each MN is recorded by its association to the access points (APs) in the WLAN. Hence, the traces represent *joint usage and mobility patterns* of the wireless network users. Using the infrastructure, one is able to capture *when* and *from where* the users access the network. All the traces, except the MIT trace, are collected from the entire campus wireless network. The MIT trace is collected from three engineering buildings in a corporate network; hence its user population is not as diverse as the other traces, and the geographic scope of trace collection is smaller. The USC trace is the only one that has coarser, per–switch port location granularity,‡ while the others have per-AP location granularity.

In this work we take smaller parts of the data sets (i.e., one-month chunks) for our analysis. We make such choices to facilitate the processing of data and make the longer traces (available for the time duration of years) comparable to shorter ones. The chosen parts are representative of the data set as a whole, and similar conclusions could be drawn if we had chosen other parts of the data set.

---

* In this work we use the terms user, node, and mobile node (MN) interchangeably. We assume that one MAC address in the trace corresponds to a unique device (MN), and a MN is always tied to the same unique user.

† However, according to the device type information provided in the Dartmouth trace archive [25], there were only 25 PDA users and 63 VoIP device users during this time period. The results we get from these small sample sizes may need further validation by studies on a larger scale.

‡ Each switch port in the USC trace has several APs connected to it. The geographic coverage of a switch port approximately corresponds to a building (or several small buildings in close vicinity) on the campus.

6.3.1 Trace Collection Methods

The methods of collecting WLAN traces can be classified into two major categories: (i) *Polling-based methods* record the association of the mobile nodes at periodic time intervals using Simple Network Management Protocol (SNMP; in the MIT trace [22]) or association tracking software on the MNs (in the UCSD trace[23]); and (ii) *Event-based methods* record MN online/offline events using logging server (e.g., syslog in Dartmouth [21] and UF trace). For the USC trace, the logs are collected from the switch (i.e., the switch creates a log when an MN associates/disassociates with one of the APs connected to the switch ports).

For traces collected using polling-based methods, we obtain only "sample points" of MN association at regular time intervals in the trace; hence the duration of association must be derived from these samples. In this work we assume that an MN is associated with the AP for four polling intervals after it is observed as associated with the AP, unless indicated otherwise by the trace (i.e., if the trace reports the MN associates with another AP, the previous association with the old AP terminates before the assumed length of four polling intervals). This approach is more robust to imperfections (e.g., record losses, wireless channel variation) in the trace collection process, but it may slightly overestimate the duration of association after an MN disassociates from an AP. However, for the results presented here, the actual difference caused by this assumption (versus other assumed association duration) is not significant.*

6.3.2 Derivation of Encounter Events from WLAN Traces

The WLAN traces provide approximate location information of individual users by their association with the access points. We further leverage this information to understand the potential opportunities of communication between the devices, assuming the infrastructure (i.e., the access points) were not present. We derive the *encounter events* between the MNs from the WLAN traces using the following assumption: The MNs can communicate with each other directly if they are associated with the same AP (or the same switch port in the USC trace). Following this assumption, the duration of the encounter events can be derived from the overlapped time intervals of association periods of different MNs with the same AP. Nodal encounters in mobile networks are important events as they provide opportunities for involved nodes to communicate directly.

We acknowledge that this assumption may not be completely accurate, as there can be scenarios where (1) MNs are associated with the same AP but are still too far apart to communicate directly, (2) MNs are able to communicate while associated with different APs, or (3) MNs may encounter each other outside the coverage of existing APs and therefore some encounter events cannot be reconstructed from the

---

* Due to space limitations, we cannot discuss this point in detail here. Please refer to [4] for more details on the comparison between polling-based and event-based trace collection methods.

WLAN association traces. However, we believe that the encounter events derived from WLAN traces capture a large number of MNs within direct communication range under current deployments of the WLAN (usually the network administrators seek to provide ubiquitous coverage when possible). In addition, the WLAN traces capture another important factor one needs to incorporate when considering interdevice communication opportunities—the usage pattern of the devices. According to our findings from the traces, many devices are not *always on*. The empirical approach of deriving encounter events from the WLAN traces has the inherent benefit of including realistic on-off usage patterns in the analysis of encounters.

6.4 Individual User Associations in Wireless LAN Traces

Among the many opportunities for potential investigation, we first focus on the following questions: How do we realistically model user behavior in campus WLANs? More specifically, if we are interested in modeling the mobility patterns of individual users in such environments, what characteristics are important to observe from the traces? And how do users in different environments differ (or not) on these aspects? We seek to answer these questions by an extensive study of WLAN traces.

6.4.1 Dimensions of Analysis

In this section, we propose metrics to *represent* different aspects of MN association behaviors in a WLAN. We shall use Figure 6.2 to illustrate. One can see the association pattern of an MN as a sequence of associated APs (shown by shades in Figure 6.2), interleaved with time periods during which the MN is *offline* (i.e., not associated with any AP). We look into four major categories to understand user behavior as follows:

a. *Activeness of users*: This category captures the tendency of a user to be online. In general, wireless network users show up in the trace intermittently, as opposed to the always-on nodes assumed in the synthetic models.

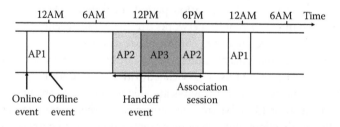

Figure 6.2 Illustration of the term definitions. Wei-jen Heu & Ahmed Helmy; "On modeling user associations in wireless LAN traces on University Campuses." IEEE WiN Mee 2006 © IEEE, with permission.

b. *Macro-level mobility of users*: This category captures how widely an MN moves in the network in the long run (i.e., for the whole trace duration) and how its online time is distributed among the APs. The intention is to capture overall long-run statistics and preference of the mobility process of the nodes.

c. *Micro-level mobility of users*: This category captures how an MN moves in the network while it remains associated with some AP (i.e., when a user "hands off" from one AP to another). The intention here is to capture the mobility of an MN while *using* the wireless network, which is different from the macro-level mobility objective.

d. *Repetitive association pattern of users*: This category captures the user association behavior with respect to time. We expect users to show repetitive structures in association patterns during similar times on different days because their mobility patterns are dictated by their daily schedule. This idea is also illustrated in Figure 6.2: the user appears at AP1 during late evenings on both days. We use the probability for a node to reappear at a previously associated AP after a certain time gap to quantify this phenomenon.

6.4.2 Activeness of the Users

Activeness of users is the first aspect we analyze in an attempt to compare the different traces. This can be represented rather straightforwardly by the *total online time fraction* of users. We define the *online time fraction* as the ratio between an MN's total online time to its *existence time*.* We plot the complimentary cumulative distribution function (CCDF)† of the online time fraction of users in various traces in Figure 6.3. We observe that in all traces, only a small portion of users have a high online time fraction, except for the Dart-04 trace. The average online time fraction is 87.68% for Dart-04 trace and between 36.44% (Dart-03) and 14.12% (UCSD) for other traces. The standard deviation for the online time fraction is large, varying from 0.24 to 0.36 for all traces. These observations argue strongly that *users have diverse on-off usage patterns, where some of the users are heavy users but many are light users.*

The distributions of the on/off times seem to depend heavily on the environments and the device types in the traces. The UCSD trace, which focused only on PDA users, is the least active one among all traces. The other traces (MIT, USC, UF, Dart-03) are not very different in online time fraction distribution. The activeness of MNs increases significantly from 2003 to 2004 in the Dartmouth trace, which agrees with the findings in [11]. Comparisons between Dart-04 with

---

* Defined as the time difference between the first online event and the last offline event of the user.
† CCDF, or the complimentary cumulative distribution function, is the probability for a random variable to exceed a given quantity x. It is a nonincreasing function taking values between the range [0,1].

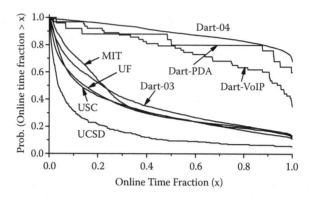

Figure 6.3   CCDF of the online time fraction. Wei-jen Heu & Ahmed Helmy; "On modeling user associations in wireless LAN traces on University Campuses." IEEE WiN Mee 2006 © IEEE, with permission.

Dart-PDA and Dart-VoIP show that during the same trace period, the handheld devices are less active than the average of the total population.

The key *characteristic* we identify regarding user on/off patterns is, although many DTN research works (or wireless mobile network research works in general) assume a futuristic environment where users are equipped with always-on mobile devices, it is still far from today's mobile device usage pattern. While omnipresent wireless mobile personal communication devices (such as PDAs or cell phones) are envisioned to emerge in the near future, it is still not clear how they are going to be used. As we observe from the traces, these devices are still not always on. Hence we argue that the usage pattern of today's mobile devices (as shown in the traces) is a reasonable starting point to investigate this futuristic scenario.

6.4.3 Macro-Level Mobility of Users

In this section, we represent the long-term mobility of users by obtaining the overall statistics of AP association history during the whole trace period. We analyze the number of APs a user associates with and the fraction of online time the user spends at each AP. Through this analysis, we gain insights about how widely an MN visits (in terms of the access points, not the actual geographical area). If a user visits more APs, and stays at more APs for a nonnegligible fraction of its online time, it is an indication that the user visits a wider range of the campus (i.e., more *mobile* in the long run) than another user who visits few APs and spends most of its online time at one or two APs.

We define the *coverage* of a user as the *percentage* of APs on the campus that the user associates with during the trace period. For the USC trace we use switch ports in place of APs. The distributions of the coverage of users in the traces are shown in

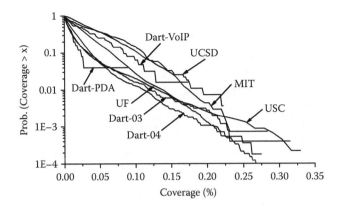

Figure 6.4 **CCDF of the coverage of users. Wei-jen Heu & Ahmed Helmy; "On modeling user associations in wireless LAN traces on University Campuses." IEEE WiN Mee 2006 © IEEE, with permission.**

Figure 6.4. We observe that *users have low coverage in all environments.* The average coverage is between 4.52% (UCSD) and 1.10% (Dart-04). None of these traces has even a single user visiting more than 35% of all APs. In the UCSD trace, the PDA users seem likely to visit a larger portion of campus than the generic users do in the other campuswide traces, due to the portability of PDAs. Similar observation applies to the VoIP devices in the Dartmouth trace, which is the most mobile subuser group in the Dartmouth trace. However, PDAs in the Dartmouth trace are less mobile than the generic users in the period we studied. We suspect that the result may be influenced by a few extreme users (there were only 25 PDA users identified during this period, and half of them visit four or fewer APs). The MIT trace is collected from only three corporate buildings; hence the relative coverage of users is a bit higher. It is important to note that the *coverage seems to remain stable with respect to time change,* although the activeness of users changes significantly (compare Dart-03 and Dart-04).

We further analyze the distribution of online time a user spends with each AP it visits. We order the APs a user ever visits during the trace period by the user's total association time with each AP, calculate the percentage of online time the user spends at each AP visited, and then average across users to get the average percentage of online time a user spends at each AP ordered by preference. The results are shown in Figure 6.5. From the figure we observe that, for all environments, the general characteristic is that *each user has very few APs at which it spends most of its online time.* In particular, for all the traces, an MN spends on average more than 65% of its online time with *one* AP and more than 95% of online time at as few as the top-5 APs combined. *The left end of the curves are similar, but the tails vary.* The higher mobility of the UCSD PDA users translates into a longer tail, where in

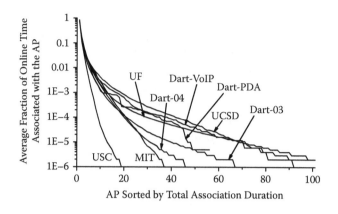

Figure 6.5 **Average fraction of time an MN is associated with APs. For each MN, the AP list is sorted based on association time before taking the average across users. Wei-jen Heu & Ahmed Helmy; "On modeling user associations in wireless LAN traces on University Campuses." IEEE WiN Mee 2006 © IEEE, with permission.**

addition to those few most-visited APs, the users also access the wireless network at many more locations with a small fraction of the user's online time as compared to other traces. Similar observations apply to the Dart-VoIP and Dart-PDA traces. It is interesting that the Dart-PDA trace shows low coverage in Figure 6.4 but a high average fraction of time associated with less-popular APs here. These two points, however, do not contradict each other. A closer investigation reveals that, although there is only a small fraction of widely visited PDAs (from Figure 6.4), those who visit many APs contribute a good part of their online time to less-popular APs. Among generic WLAN devices, the users in the UF trace spend higher fractions of time at more locations than other campus wireless network users.

6.4.4 Micro-Level Mobility of Users

In this section, we analyze the per-association session mobility of a user, which reflects its short-term mobility. This represents a different dimension of user mobility as compared to the previous section: How mobile the user is while *using* the network. We use handoff statistics as a measure of user mobility while using the network. However, we discover that a lot of handoff events are due to the so-called *ping-pong effect* rather than real movements. The term ping-pong effect refers to the phenomenon of excessive handoff events due to disturbance in wireless channels while the MN itself might be stationary. Hence, we cannot directly link the handoff statistics to the micro-level mobility of the users. Development of better filters for ping-pong effects is needed before we can really understand the micro-level mobility from the WLAN traces.

Our investigation shows that the number of handoff events depends heavily on the device type (e.g., VoIP devices have the most handoff events because they remain on while moving around in the network) and the environment (i.e., how the access points are deployed, etc.). On the other hand, there is no clear relationship between the duration of an association session and the number of handoff events within the session. In some cases, we see extremely long sessions without any handoff events or an extremely high number of handoff events in a session with short duration. The correlation coefficients between session lengths and handoff counts for all the studied traces are between 0.377 and 0.030. The session length and the handoff count have a weak linear correlation to each other in all traces.

We further analyze whether the sessions with high handoff counts are all from a small set of extremely mobile users. For each user, we calculate the average number of handoff events per unit time (i.e., the *handoff rate*) for each of its sessions and then calculate the mean and variance for the user's handoff rate from all sessions of this user. If a high degree of micro-level mobility leading to the high handoff count is an intrinsic property for some users, we should see that those users show high average and low variance in their handoff rates. We use the coefficient of variation (the standard deviation divided by the mean) to understand the degree of variation in the handoff rates for users. In Figure 6.6, we show the CDF of the coefficient of variation of the handoff rate for the studied traces. Only the users with more than one session and one handoff event are considered in the graph, since users with only one session automatically result in 0 variance for the handoff rate. From the figure, we see that *the handoff rate displays high variance for most of the users.* In all traces, more than 60% of users have a coefficient of

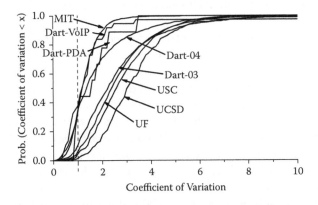

Figure 6.6   CDF for the coefficient of variation of the handoff rate of the users. Note that for all traces the coefficient of variation is larger than 1.0 for at least 60% of the MNs with more than one session. Wei-jen Heu & Ahmed Helmy; "On modeling user associations in wireless LAN traces on University Campuses." IEEE WiN Mee 2006 © IEEE, with permission.

variation of the handoff rate larger than 1.0 (i.e., the standard deviation is larger than the mean). This indicates that, even for a given MN, the handoff rate varies drastically from session to session.

Combining the observations in the preceding paragraphs, we conclude that handoff events not only distribute unevenly among users but also happen unevenly among the association sessions for the same user. This indicates that the handoff events are greatly influenced by the environmental condition when a session is established rather than the property of the MN that initiates the session. The reduction of the ping-pong effect is thus an important issue to consider in making better interpretations of the micro-level user mobility from the WLAN traces and warrants further study.

6.4.5 Repetitive Association Pattern of Users

Naturally, user behavior changes with respect to time of the day and day of the week, as people follow daily and weekly schedules in their lives. In some cases, the user association pattern repeats itself day to day or week to week. In this section, we represent such repetitive patterns by the probability of users to reappear at (i.e., reassociate with) the same access point after some time gap. To quantify this probability for a given time gap (for example, one day), we sample the locations of a given user throughout the trace period and count the percentage of sample pairs that are separated by a one-day time difference, and the user indeed appears at the same location (i.e., when the user is online at the same AP in both samples). This is the empirical probability for this user to reappear at the same location after a time gap of one day. We then take an average of the numbers from all users to obtain the overall likelihood of users to reappear at the same location after a given time gap.

In Figure 6.7 we show the *reappearance probabilities* for all the traces. We see that in most of these traces (i.e., USC, MIT, Dart-03, Dart-04) we observe noticeably higher reappearance probability if the time gap is close to integer multiples of one day. This characteristic indicates that *users have the strongest tendency to show repetitive association pattern at the same time of each day*. It is also interesting to observe that, for these traces, the *reappearance probability for the time gap of seven days (i.e., one week) is the second highest,* only slightly lower than that for the time gap of one day. This indicates that the weekly repetitive pattern is also strong in these traces. On the other hand, the UCSD and UF traces show little repetitive pattern as there are almost no obvious spikes in its reappearance probability curve. For UCSD, this can be attributed to its user population being PDA users. Unlike laptops, which are more related to work, PDAs are usually used in a more casual way in short, scattered time periods. Hence it is expected that PDA users show less regularity in their usage pattern. For the UF trace, we only observe a slight (almost unnoticeable) peak at the seven-day time gap.

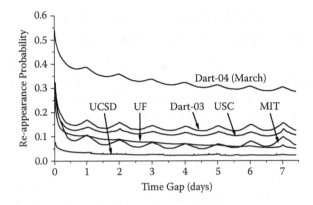

Figure 6.7 **User reappearance probability at the same location versus time gap. The peaks represent time gaps after which the users are more likely to reappear at their current locations. Wei-jen Heu & Ahmed Helmy; "On modeling user associations in wireless LAN traces on University Campuses." IEEE WiN Mee 2006 © IEEE, with permission.**

6.4.6 Conclusions for the Individual Behavior Analysis

To summarize, the findings from the traces point out important common features in all studied environments. Wireless network users in university campuses and in corporate networks are characterized by (1) a limited number of visited APs in the network and a large proportion of online time spent at very few of its most-visited APs. The coverage of users never exceeds 35% in all traces, and users spend more than 95% of their online time with as few as *five* APs. Furthermore, these numbers seem to remain relatively stable in a given environment, even if the WLAN gains popularity and users become more active. (2) Periodic association patterns with recurrent daily/weekly location visits. (3) Large percentages of offline time. Except for the Dart-04 trace, fewer than 20% of users are always on, and more than 68% of users are offline more than 50% of time. Even in the most active Dart-04 trace, more than 30% of users are not always on. We believe that these metrics capture important characteristics about users in wireless networks that are largely overlooked by earlier work on mobility modeling and wireless network simulation. They clearly point out that conventional mobility models (e.g., random waypoint, random walk, etc.), in which nodes are always on and move about the whole simulation area with equal likelihood without any spatial-temporal dependency, are inadequate for describing realistic user behaviors in complex environments like university campuses. Based on these empirical characteristics, one can design more realistic mobility models. For example, see [6] for the *time-variant community model* we propose to capture these observed mobility features.

6.5 Nodal Encounter Patterns and Their Impact on Opportunistic Networks

Our work in the previous section provides a good understanding of the WLAN user mobility process. However, although the understanding of individual behavior is important in itself, it does not reveal how MNs interact with one another in the real traces. In this section we go beyond the level of individual users and take a more macroscopic view. We consider an important event between mobile nodes in wireless networks—*encounters*. Nodal encounters are frequently used in opportunistic networks as the major opportunities for direct nodal communication. We again use month-long WLAN traces from university and corporation campuses in this chapter (i.e., the MIT, Dart-03, Dart-04, UCSD, USC, UF traces from Table 6.1). We seek to understand encounters derived from the traces from a different perspective—we take a holistic view of all encounter events happening between all the nodes in the network and study the global encounter patterns with a graph analysis approach. Such an analysis sheds light on the diverse, nonhomogenous nature of users in the given environments in terms of their encounter events with other nodes. Furthermore, we evaluate the feasibility of forming an infrastructureless network capable of reaching most of the nodes through time-varying, partial connectivity through encounters.

6.5.1 Dimensions of Analysis

Again, in this section, we consider different metrics and representations to understand various aspects of the user encounter pattern. We briefly outline each representation and the questions we try to answer below:

a. *Distribution of encounter events*: The most straightforward approach to understanding encounter events is by counting their occurrences. In particular, how many other unique nodes does a given node meet with? How many total encounter events are there for each node? Does the encounter event count follow a certain distribution? Answering these questions allows us to have a first-level understanding of the encounter events and, surprisingly, the encounter events derived from the traces are qualitatively different from encounter events found in synthetic models.

b. *Encounter-relationship graph*: We capture the potential of encounter events to link individual nodes into a connected network by representing these events with a graph. In this encounter-relationship graph (ER graph), two nodes are connected by a link if they ever encounter each other during the studied time period. We then analyze various graph metrics of the ER graph and their evolution as the studied time period changes, to understand a special pattern formed by these mobile nodes meeting each other.

c. *User friendship*: The next question we ask about interuser relationship is whether one can identify *close* relationships between certain pairs of users from the trace. We propose several metrics to quantify user relationships based on encounter duration, encounter count, or locations at which users encounter each other. We observe the distribution of such relationships among all user pairs and also analyze how the structure of ER graphs changes with respect to selective addition of links based on a different degree of interuser friendship.

d. *Information diffusion*: Finally, we propose information diffusion experiments to understand how messages could be spread among users *without* the help of the infrastructure. We use a simple message-spreading strategy (i.e., epidemic routing [18]) to investigate whether it is possible to rely on mutual encounters to spread messages across large-scale mobile networks. We first perform simulations under ideal scenarios and then investigate the richness of encounter events by performing more simulations after the removal of some nodes or encounter events.

6.5.2 Distribution of Encounter Events

The distribution of the encounter events is the first step in understanding the structure of inter-MN relationships in the traces. The first representation we choose is the percentage of other nodes each MN encounters through the whole trace period (one month). Figure 6.8 shows the CCDF of this metric, and we observe that all the nodes in WLAN traces encounter only at most about 50% of the user population within a month, with the UCSD trace being the only exception. This may be partly due to the fact that the 275 PDA users in the UCSD trace were all selected from the freshman class, and they tend to stay in several common dorms as stated in [10] (in other words, the MNs in this trace are selected from a *correlated subgroup* of the whole population on campus). In all other traces, on average, an MN encounters only 1.33% (UF) to 6.70% (Dart-04) of the whole user population within the

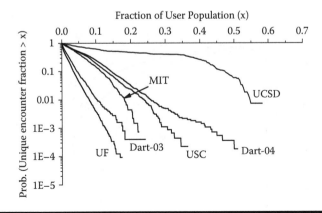

Figure 6.8   CCDF of the unique encounter fraction. Wei-jen Heu & Ahmed Helmy; "On Nodal Encounter Patterns in Wireless LAN Traces." IEEE WiN Mee 2006 © IEEE, with permission.

30-day trace period. The small average encounter ratio is a combined result of several reasons: (1) most MNs are not always on (cf., Section 6.4.2), and (2) most MNs do not visit many APs (cf., Section 6.4.3); hence, they can only meet with those who also visit this small set of APs.

Such low encounter percentages shown in the traces is a very distinct characteristic not observed in any of the *i.i.d. random mobility models*, such as random waypoint or random walk, which are widely used for performance evaluation in the literature [13]. As shown in [14], the unique encounter fraction reaches 100% for all nodes within a short period of time under these models. This is because in typical i.i.d. random mobility models, all nodes follow the same model to make movement decisions, albeit with randomness, and eventually encounter all other nodes. The encounter pattern from real wireless network traces, on the other hand, reflects that the university campus is a heterogeneous environment rather than a homogeneous one. Thus, to better understand how protocols perform in such a complex environment, using homogeneous synthetic models is not sufficient.

We also represent the encounter events by the CCDF of the total encounter event count an MN has throughout the trace period in Figure 6.9. We observe the total encounter counts for MNs in each trace span across several orders of magnitude. There are both MNs with extremely few or many encounters. This is again evidence of heterogeneous encounter characteristics among MNs. The actual number of total encounters depends on the size of the populations in the traces. Large traces (i.e., the USC, UF, and Dartmouth traces) tend to have more encounters than small traces (i.e., the UCSD trace). However, regardless of the size of population, the curves for the total encounter count derived from WLAN traces seem to follow the *BiPareto distribution*. We fit the BiPareto distribution curves to the empirical distribution curves, and use the Kolmogorov-Smirnov test [16] to

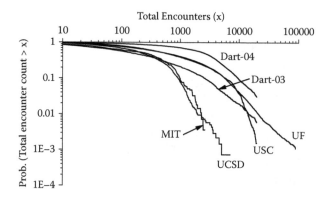

Figure 6.9 CCDF of the total encounter count. Wei-jen Heu & Ahmed Helmy; "On Nodal Encounter Patterns in Wireless LAN Traces." IEEE WiN Mee 2006 © IEEE, with permission.

examine the quality of fit. The resulting D-statistics for all traces are between 0.068 and 0.025, which indicates that we have a reasonably good fit between the BiPareto distribution curves and the empirical distribution curves.*

A closer investigation of the relationship between the unique encounter count and the total encounter count of the same MN reveals that a *high unique encounter count does not always imply a high total encounter count.* The correlation coefficients between the unique encounter count and the total encounter count for various traces range from 0.732 to 0.195. Except for the UCSD trace, all other traces have correlation coefficients below 0.6. Most prominently, we observe that some node pairs do not have many unique encounter counts but show high total encounter counts. This indicates that some node pairs may have a lot of repetitive encounters, suggesting that these node pairs have closer relationships than other pairs. This point warrants further study, and we will show some initial attempts to quantify the friendship between MNs in Section 6.5.4.

6.5.3 Encounter-Relationship Graph

In Section 6.5.2, we saw that MNs have a low percentage of unique encounters. Given this fact, we raise a question regarding the possibility of establishing campuswide relationships among the majority of MNs via encounters alone. That is, do encounters link MNs on the campus into one single community, or just many small cliques?

To investigate this question, we define a graphical representation of the encounter events, the *static encounter-relationship graph (ER graph)*, as follows: Each MN is represented by a node in the ER graph, and an edge is added between two nodes if the two corresponding MNs have encountered each other at least once during the studied trace period. By the construction of the ER graph, we collect all encounter events between MNs within a time period and collapse them on a static graph. The exact timing of encounters is ignored, and we focus on the structure of interconnections built between nodes by available encounter events during that period of time. In other words, the concept of the ER graph is introduced to capture the potential of establishing a connected network among MNs based on direct encounters alone and to understand the structure of such a network.

We use three important metrics to describe the characteristics of the encounter-relationship graphs, defined as follows:

The clustering coefficient (CC) is used to describe the tendency of nodes to form cliques in a graph. It is formally defined as [17]:

$$CC = \frac{\sum_{n=1}^{M} CC(n)}{M},$$ (6.1)

---

* Due to space limitations, the details about the Kolmogorov-Smirnov test and the parameters of the fitted BiPareto distribution curves are not listed here. Details can be found in [5].

where

$$CC(n) = \frac{\sum_{a,b \in N(n)} I(a \in N(b))}{|N(n)| \cdot (|N(n)| - 1)}. \tag{6.2}$$

$N(n)$ is the set of neighbors of node n in the ER graph and $|N(n)|$ is its cardinality. $I(\cdot)$ is the indicator function. M is the total number of nodes in the graph.

Intuitively, the clustering coefficient is the average ratio of neighbors of a given node that are also neighbors of one another. Higher CC indicates a higher tendency that neighbors of a given node are also neighbors to each other, or heavy cliquishness in the relationship between MNs formed through encounters.

The **disconnected ratio (DR)** is used to describe the connectivity of the ER graph. It is defined as:

$$DR = \frac{\sum_{a=1}^{M} (M - |C(a)|)}{M(M-1)}, \tag{6.3}$$

where $C(a)$ is the set of nodes that are in the same connected subgraph with node a. DR indicates, on average, the percentage of unreachable nodes starting from a given node in the graph.

The **average path length (PL)** is used to describe the degree of separation of nodes in the ER graph. It is defined as:

$$PL = (1 - DR) \cdot PL_{con} + DR \cdot PL_{disc}, \tag{6.4}$$

where PL_{con} is the average path length among the connected part of the ER graph, defined as:

$$PL_{con} = \frac{\sum_{a=1}^{M} \sum_{b \in C(a)} PL(a,b)}{\sum_{a=1}^{M} |C(a)|}. \tag{6.5}$$

$PL(a, b)$ is the hop count of the shortest path between node pair (a, b) in the ER graph. PL_{disc} is the penalty on the average path length for *disconnected* node pairs in the ER graph. We use the average path length of the regular graphs (defined later) with the same node number and average node degree for PL_{disc}.

We analyze how the above metrics evolve for the ER graphs derived from various studied periods of WLAN traces. Taking the USC trace, the Dartmouth trace (Dart-04), and the UCSD trace as examples, we show the evolution of the three metrics with respect to various studied trace periods in Figure 6.10. The graphs for

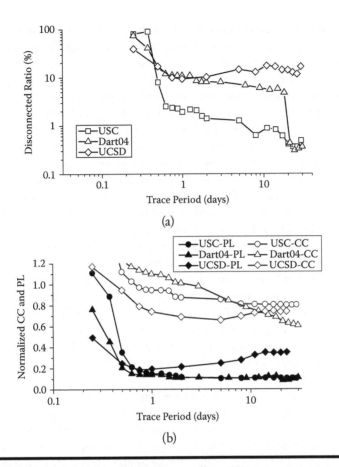

(a)

(b)

Figure 6.10 Change in the ER graph metrics with respect to trace period. (a) Disconnected ratio. (b) Normalized clustering coefficient and average path length. The figure is cut from above to show the details between 0 and 1 on the y-axis. Wei-jen Heu & Ahmed Helmy; "On Nodal Encounter Patterns in Wireless LAN Traces." IEEE WiN Mee 2006 © IEEE, with permission.

other traces show very similar trends and are not shown here due to space limitations (please refer to [5] if interested).

From Figure 6.10 (a) we note that given sufficient long trace durations, the ER graphs have low *DR* (not larger than 10% for traces longer than one day in most cases), which implies that *nodal encounters are sufficient to provide opportunities to connect almost all nodes in a single community*, even though each node encounters only a small subset of MNs directly. This is an encouraging result that points out the feasibility of building a large, wide-reach network relying only on direct encounters. Although the *DR* starts out very high with very short trace periods (i.e., for trace durations under one day) since MNs have not moved around to create encounters yet, it decreases rather quickly as the trace period increases. Within one day, the *DR* for each trace reduces to around 10%. Although the numbers of MNs in the ER graph keep increasing as we look at longer trace periods, in most cases the *DR* does not change significantly after one day.

Another interesting characteristic of the ER graph is revealed by the other two metrics, the clustering coefficient (*CC*) and the average path length (*PL*). To highlight a unique property of these ER graphs, we also calculate the *CC* and the *PL* for *regular graphs* and *random graphs* with the same corresponding total node number *M* and average node degree *d*. These quantities can be calculated according to the equations in Table 6.2. In the regular graphs, nodes are first arranged on a circle and each node is connected to *d* closest neighbors on the circle. In the random graphs, *d* randomly chosen nodes are assigned as neighbors for each node. Typically, regular graphs have high *CC* and *PL* while random graphs have low *CC* and *PL*. They are the two extreme cases on the spectrum. In Figure 6.10 (b), we show the normalized *CC*s and *PL*s of the ER graphs for various trace periods. These normalized metrics represent, on a scale from 0 (corresponding to the random graph) to 1 (corresponding to the regular graph), where the metrics of the ER graphs fall. They are defined as: $CC_{norm} = (CC - CC_{rand})/(CC_{reg} - CC_{rand})$, $PL_{norm} = (PL - PL_{rand})/(PL_{reg} - PL_{rand})$, where CC_{norm} and PL_{norm} represent the normalized *CC* and *PL*, respectively. The subscripts *reg* and *rand* imply that the corresponding metric is obtained from the regular graph and the random graph, respectively, with the same total node number and average node degree.

We observe that ER graphs display high normalized *CC*s, which are close to those of the corresponding regular graphs (i.e., normalized *CC*s being close to 1, and in some cases even higher than 1), and low normalized *PL*s, which are close to those of the corresponding random graphs. This highlights that a special pattern of encounters exists in all WLAN traces: nodes visiting similar sets of APs are highly likely to encounter all others and introduce highly connected clusters among these nodes, leading to high *CC*. This phenomenon is especially obvious for very short traces, since most MNs do not change their association to the APs to create many encounters. The ER graphs for short trace periods feature many small disconnected cliques, each of them being a full-mesh formed by MNs associated with the same AP for that trace period. As we look at longer traces, some of the nodes in one cluster also have random

Table 6.2 Equations for the *CC* and *PL* for the Regular and Random Graphs with *M* Nodes and Average Node Degree *d*

Graph Type	Clustering Coefficient	Average Path Length
Regular graph	$3(d-2)/4(d-1)$	$M/2d$
Random graph	d/M	$\log(d)/\log(M)$

Note: See [12] and [17].

encounters with nodes in other clusters, and these links serve as the shortcuts in the ER graphs that reduce the *PL*. In previous literature, graphs with high *CC* close to the regular graphs and low *PL* close to the random graphs are known as *Small World graphs* [12], [17]. By looking at the ER graph characteristics of various traces, we indicate that the ER graphs formed by encounters among wireless network users appear to be Small World graphs. We also observe that both *PL* and *CC* converges to its final values rather quickly in about one day for most traces, although the size of ER graphs keeps increasing as more nodes appear in longer traces.

6.5.4 Capturing User Friendship in WLAN Traces

In our daily lives, we are bound to meet with colleagues and friends much more often than others. In this section, we try to investigate, using the traces, whether such an uneven distribution of closeness among MN pairs exists, and we try to measure it using the concept of *friendship dimensions* and understand its influences on the ER graphs when we include friends with different degrees of closeness in the graph. We propose to represent friendship between MN pairs based on three different dimensions—encounter duration, encounter count, and encounter AP count, with the definitions below. The intuition behind the definitions is simple—the longer, the more frequent, and at more locations two nodes meet, the closer the relationship is between them, usually. However, notice this "friendship" derived from the trace may or may not reflect actual social friendship, which is impossible to validate from anonymous traces.

- **Friendship based on encounter time** is defined as $Frd_t(a, b) = E_t(a, b)/OT(a)$, which is the ratio of the sum of encounter durations between node a and b, $E_t(a, b)$, to the total online time of node a, $OT(a)$. This is an index for how close node b is to node a based on the duration of encounters. Note that in general $Frd_t(a, b) \neq Frd_t(b, a)$ and $0.0 \leq Frd_t(a, b) \leq 1.0$ for any node pair a and b.
- **Friendship based on encounter count** is defined as $Frd_c(a, b) = E_c(a, b)/S(a)$, the ratio between the count of association sessions of node a that contain encounter events with node b, $E_c(a, b)$, to the total association session count of node a, $S(a)$.
- **Friendship based on encounter AP count** is defined as $Frd_{AP}(a, b) = E_{AP}(a, b)/AP(a)$, the ratio between the number of APs at which node a has encounter events with b, $E_{AP}(a, b)$, to the total APs node a visits, $AP(a)$.

We first analyze how friendship indexes distribute among all node pairs in the traces. As shown in Figure 6.11, the CCDF curves of friendship indexes based on encounter time seem to follow exponential distributions for all campuses. We again use the Kolmogorov-Smirnov test [16] to examine the quality of fit. The resulting D-statistics for all traces are between 0.0356 and 0.0052, which indicates that we have a reasonably good fit between the exponential distribution curves and

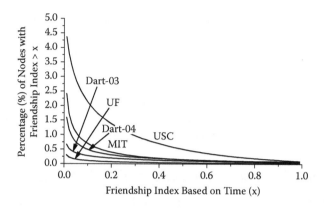

Figure 6.11   CCDF of the friendship index based on time. Wei-jen Heu & Ahmed Helmy; "On Nodal Encounter Patterns in Wireless LAN Traces." IEEE WiN Mee 2006 © IEEE, with permission.

the empirical distribution curves. This exponential distribution of the friendship indexes is an indication that the majority of nodes do not have a close relationship with one another. In all the traces, only less than 5% of ordered node pairs (*a*, *b*) have a friendship index $Frd_t(a, b)$ larger than 0.01. Among all node pairs with a nonzero friendship index, only 4.47% of them have a friendship index larger than 0.7, and another 11.85% of them have a friendship index between 0.4 to 0.7. In other words, we can say that the friendship between the MNs is very "sparse" (i.e., only a few pairs of nodes show a strong relationship based on the above definitions). Friendship indexes based on encounter frequency or encounter AP count also show similar sparse characteristics (i.e., following exponential distributions).

We also look into the issue of whether the friendship is symmetric. We calculate the correlation coefficients between $Frd_t(a, b)$ and $Frd_t(b, a)$ for all the traces, as well as the friendship indexes based on other definitions. The correlation coefficients are low in most cases (ranging from 0.415 to –0.024, the only exception being 0.629 for a friendship index based on encounter time for Dartmouth 2004 trace), implying high asymmetry in friendship indexes[*]—when node *a* is a good friend of node *b*, the reverse may not be true.

After seeing the sparseness and high asymmetry of the friendship between the MNs, we ask the following question: If we consider friendship when establishing relationships between nodes, how would that influence the structure of the encounter-relationship graphs? Typically, an MN may not trust any random MN it encounters but is more likely to maintain connections selectively only with those MNs that are considered trustworthy. For example, an MN may choose to trust those MNs with which it has high friendship indexes (e.g., it encounters frequently).

---

[*] Due to space limitations, please refer to [7] for detailed numbers.

The criteria of choosing the nodes with which to keep a relationship may influence the structure of the ER graphs. We thus devise an experiment to include friends with various friendship indexes in the ER graph and see how it influences the structure of the graph. We use the friendship index based on time as an example below.

For each node (say node *a*), we sort the list of all nodes it encounters according to the friendship index, $Frd_t(a, b)$, for all *b* where $Frd_t(a, b) \neq 0$. After sorting, each node picks a certain percentage of nodes from the list with which to establish a link on the ER graph. Note that the links in these ER graphs are directed links when we consider friendship, because friendship is asymmetric between a given node pair. Therefore, we replace the definition of the clustering coefficient of a node in Equation (6.2) with the following

$$CC(n) = \frac{\sum_{a \in F(n)} \sum_{b \in F(n)} I(a \in F(b))}{|F(n)| \cdot (|F(n)| - 1)}, \tag{6.6}$$

where $F(n)$ is the set of chosen friends with which node *n* will maintain links. Note that $b \in F(a)$ does not imply $a \in F(b)$, and vice versa. Intuitively, here the clustering coefficient is the average ratio of the included friends of a node that also include each other as a friend. When calculating the average path length and the disconnection ratio, we follow the same definitions as introduced in Section 6.5.3, but the paths must follow the direction of edges on the ER graph.

Following the above definitions, we obtain the graph metrics when each node includes certain percentages of encountered nodes from the top, middle, or bottom of the sorted node list according to the friendship index based on time. The figures are shown in Figure 6.12. We use the USC trace as an example, and similar results are also observed in other traces. The figures show a clear trend that if neighbors ranked high in the friendship index are included, the resulting ER graph shows stronger clustering and the average path length is much higher. In other words, it is more inclined toward a *regular graph*. The result stems from the fact that top friends of a given node are also likely to be top friends between one another, forming small cliques in the graph. The clustering coefficient remains high due to these cliques. The disconnected ratio and the average path lengths are high due to the lack of links between different cliques. On the other hand, when low-ranked friends are included in the graph, the links included are distributed in a more random fashion, which is reflected by the low clustering coefficient and low average path length, and the result is more similar to a *random graph*. As a larger portion of friends are included in the graph, all three metrics converge to the values when all encounters are included.*

---

* Note that including 100% of friends means to include every MN encountered in the ER graph. Hence, the resulting ER graph is the same as the one defined earlier in Section 6.5.3.

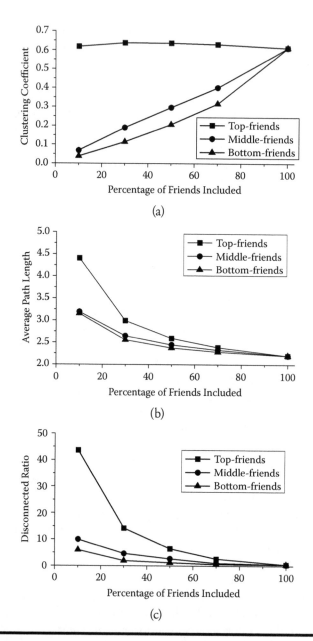

Figure 6.12 Metrics of the encounter-relationship graph by taking various percentages of friends. (a) Clustering coefficient. (b) Average path length. (c) Disconnected ratio. Wei-jen Heu & Ahmed Helmy; "On Nodal Encounter Patterns in Wireless LAN Traces." IEEE WiN Mee 2006 © IEEE, with permission.

Therefore, although it is possible to create a campuswide community based solely on nodal encounters, it is not sufficient to trust and utilize only top-ranked friends (or the MNs one encounters frequently), as this results in an ER graph with high clustering coefficient and average path length and may lead to a disconnected network. Similar to social networks, close friends in wireless networks often form cliques. In order to remain connected to a larger community, one should also use some randomly chosen users (or middle-ranked friends), as they are the key to reducing the degree of separation in the underlying ER graph.

6.5.5 Information Diffusion Using Encounters

In addition to establishing relationships between nodes, encounters can also be utilized to diffuse information throughout the network. In the opportunistic network model, information is spread with nodal mobility and encounters. The speed and reachability of information diffusion among the nodes are determined by the actual pattern and sequences of encounter events, and this is not captured by the static ER graph representation discussed above. In this section, we simulate a simple message forwarding protocol (i.e., epidemic routing [18]) on top of the encounter events derived from the traces to answer the following questions: Are the *current* encounter patterns between MNs in wireless networks rich enough to be utilized for information diffusion? If the answer is yes, how robust is it to imperfection in the environment, such as bandwidth limitation or lack of incentive (i.e., some users being selfish)? In this section, we first analyze the optimistic expectation of the potential performance of information diffusion under idealistic assumptions and then remove some of the assumptions and evaluate the performance in more realistic settings.

6.5.5.1 Ideal Scenarios

As the first step in understanding the potential of information diffusion under realistic encounter patterns, we make the following idealistic assumptions: (1) There is sufficient bandwidth and reliable communication between MNs and sufficient storage space on all MNs, (2) MNs discover the communication opportunities immediately when they encounter other MNs, and (3) every MN in the network is willing to participate in forwarding information for others. In this experiment, we focus more on analyzing how the encounter pattern itself influences the performance of information diffusion. The experiments in the following subsections deal with more realistic scenarios when some of the above assumptions are removed.

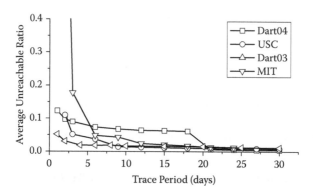

Figure 6.13   Unreachable ratio of information diffusion using epidemic routing. Wei-jen Heu & Ahmed Helmy; "On Nodal Encounter Patterns in Wireless LAN Traces." IEEE WiN Mee 2006 © IEEE, with permission.

The diffusion mechanism we use is the following: When a source node has a message to send, it simply transmits it to all nodes it encounters if they have not received the information yet. All intermediate nodes cooperate in the information diffusion process, keeping a copy of received information and forwarding it the same way as the source node. This simple approach is known as *epidemic routing* in the literature [18]. In a perfect environment with sufficient resources, it achieves the lowest delay and the highest delivery rate possible.

In all the simulations in this section, we use a traffic pattern in which the source node has some information to send to all other nodes. The source starts to diffuse the information when it is first online. As time evolves, nodes encounter each other and an increasing portion of the whole population receives the information. We study the percentage of nodes that have received the information within various trace periods and show the results in Figure 6.13, using the USC, Dart-04, Dart-03, and MIT traces as examples. Each point in the figures in this section is an average value of sending messages from 30% of the nodes that appear earliest in the corresponding trace period.

From Figure 6.13 we observe that even within a short trace period (e.g., two days) the information can reach a moderate portion of the population because the unreachable ratio is less than 25% in all traces. As the trace period increases, reachability also improves. In all except the Dart-03 trace, the unreachable ratios are less than 2% if we allow one month for information diffusion. Given that most nodes encounter only a very small portion of the whole population (Figure 6.8), this result is beyond our original expectation. *It gives a positive confirmation that it is potentially possible to deliver information relying only on encounters* in a campus environment, with a high success rate and under *current* user behavior patterns.

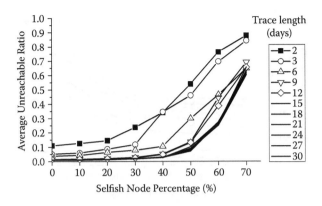

Figure 6.14   USC trace: Unreachable ratio with various selfish node percentages and trace periods. Wei-jen Heu & Ahmed Helmy; "On Nodal Encounter Patterns in Wireless LAN Traces." IEEE WiN Mee 2006 © IEEE, with permission.

6.5.5.2 Selfish Users

After studying the ideal case, we consider a more realistic setup. We first relax the ideal assumption (3) above. In some cases, some nodes may not be cooperative in propagating the information. To understand how uncooperative users potentially influence the feasibility of information diffusion, we carry out the following experiment—we make a portion of users *selfish* such that they never forward information for other sources, and we study the performance degradation under this setup. For each of the trace periods used, we increasingly make a certain percentage of nodes selfish, starting with those with the *highest unique encounter counts*. By making nodes with high unique encounter counts selfish first, we eliminate more transmission opportunities than if we pick selfish nodes randomly; hence, we expect to observe a greater impact on performance.

The relationship between the percentage of selfish nodes and the unreachable ratio for the USC trace is shown in Figure 6.14.* The result is very surprising. For all trace periods tested, the unreachable ratio does not increase significantly until at least 20% of the nodes are selfish. The performance is even more robust if we take a longer trace period. This implies that even if a significant portion of users are not willing to propagate information for others, the underlying nodal encounter pattern is rich enough for the information to find an alternative way through. Hence, the delivery rate is quite robust for up to an intermediate percentage of selfish nodes. Note that we make the MNs with most unique encounters selfish first, and hence

---

* Due to space limitations, we use only the USC trace to illustrate the point here. More graphs on results from other traces can be found in [5] and [7].

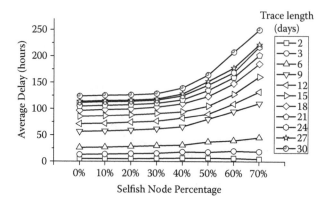

Figure 6.15 USC trace: Average message delay with various selfish node percentages and trace periods. Wei-jen Heu & Ahmed Helmy; "On Nodal Encounter Patterns in Wireless LAN Traces." IEEE WiN Mee 2006 © IEEE, with permission.

the performance of information diffusion is robust even if the nodes with the *most* chances to propagate the information are not cooperative. We further show how the average delay of information diffusion changes with the increasing selfish node percentage in Figure 6.15 for the USC trace. In the figure, for all tested trace durations, the average delay does not increase significantly until more than 40% of the nodes are selfish. This implies that the average delay is also robust against selfish user behavior up to a certain degree.

6.5.5.3 Removal of Short Encounters

Another idealistic assumption we made is that the MNs can communicate with each other successfully regardless of the duration of encounter events. This may not be true in realistic scenarios due to wireless bandwidth limitations or delay in discovering encounter events. To address this issue, we remove short-lived encounter events that do not permit prompt discovery and useful information exchange in the following experiment and reevaluate the performance of information diffusion with different minimum duration thresholds for an encounter event to be considered useable.

In Figure 6.16, we show the relationship between the unreachable ratio versus the lower limit of encounter duration (i.e., we remove all encounter events that have shorter durations than the value), using the first 15-day traces from USC and Dartmouth as examples. From the graph we observe that the unreachable ratio increases almost linearly as we increase the lower limit of usable encounter duration. There is no obvious point at which the performance suddenly degrades severely. We carry out the experiments up to the shortest usable encounter threshold set at *one hour*, a rather demanding scenario. Even in such cases, the unreachable ratio is below 30%. This implies that removing encounters of short duration does not cause abrupt degradation in the

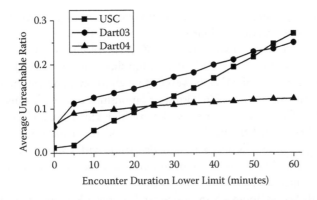

Figure 6.16 The unreachable ratio after removing short encounters below the duration lower limit. Wei-jen Heu & Ahmed Helmy; "On Nodal Encounter Patterns in Wireless LAN Traces." IEEE WiN Mee 2006 © IEEE, with permission.

performance of information diffusion in terms of both the reachability and the average delay (see Figure 6.17). In other words, *short encounter events are not the key reason for the success of information diffusion.* The encounter events with long durations are also rich enough to be utilized for message propagation in most cases.

6.5.6 Conclusions for the Encounter Analysis

In this section, we investigated the encounters between MNs in WLAN traces from five sources. We found that MNs encounter only a small subset of other nodes (on average between 1.33% to 6.70%), and the total encounter counts follow the BiPareto

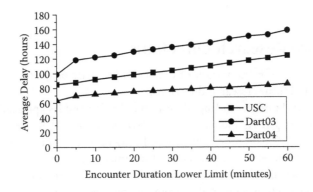

Figure 6.17 The delay after removing short encounters below the duration lower limit. Wei-jen Heu & Ahmed Helmy; "On Nodal Encounter Patterns in Wireless LAN Traces." IEEE WiN Mee 2006 © IEEE, with permission.

distribution. In spite of a low percentage of unique encounters, the relationship graph constructed using encounters alone connects most of the MNs. Furthermore, such encounter-relationship graphs display Small World graph characteristics, and its graph properties converge to its long-term value within short time periods. The relationship between different pairs of MNs, however, is much skewed and can be modeled by the exponential distribution. Establishing relationships only with those considered as high-ranked friends leads to a network with high clustering and disconnections, while using low-ranked friends is the key for good reachability in the encounter-relationship graphs. Finally, using a simulation study with a simple protocol, we also display the potential for information diffusion without relying on the infrastructure, utilizing encounters and mobility of MNs alone.

The Small World approach to understanding the ER graphs and the result of information diffusion experiments both highlight the positive potential of building a campuswide network without infrastructures. The robustness of information diffusion brings up two interesting points:

1. For message delivery, the delivery ratio and delay are not affected significantly, even if we cannot choose the shortest paths due to noncooperative users or unutilized short encounters.
2. On the other hand, it would be difficult to prevent diffusion of harmful or malicious messages, such as computer worms or viruses, from propagating through encounters [19].

Both observations are due to the richness in the underlying encounter pattern providing abundant chances for message delivery. Small World encounter relationship patterns can be considered as an ambient structure in human networks and can be used to design efficient message forwarding protocols. This is our major future direction, and some preliminary results are available in [8].

6.6 Future Research Directions

Based on the TRACE framework of understanding user behaviors, we point out several potential future research directions:

Fundamental understanding of human behavior and its application in protocol/service design. As mobile network devices become ubiquitous and tightly coupled with individuals, monitoring users through collected traces provides a powerful platform to observe and understand fundamental human behavior. Although the central focus in the chapter is about the lessons learned on mobility and encounter patterns, the characteristics identified from the traces really have generic utilizations beyond the scope of the discoveries. One prominent direction is to build *behavior-aware* protocols or services, where the user characteristics can be leveraged to introduce more efficient network protocols or services geared toward user preferences.

User privacy preservation. As users spend an increasing amount of time and perform more tasks online, users are more exposed to the danger of privacy leaks through monitoring. While trace-based studies show great promise, it is also of utmost importance to defend user privacy. Issues related to privacy emerge at all levels in such a project, including in the collection and postprocessing of traces (better techniques should be devised to keep information anonymous) and the design of behavior-based message dissemination protocols (users should be able to decide when and how much to reveal their behavior profiles, or opt out altogether). The mobile network paradigm provides various new challenges in user privacy.

6.7 Conclusions and Our Contributions

In this chapter we outlined the TRACE framework, a step-by-step procedure to investigate user behaviors based on realistic data sets, identify important characteristics, and understand their impact on opportunistic networks. We showed the application of the TRACE framework at two different levels, *individual user associations* and *global encounter patterns*, to display the applicability of the framework.

The major contributions of this analysis are the following: First, by using WLAN traces from five different sources, comparing the results, and highlighting both similarities and differences, it is the largest-scale trace-based study in the literature to our knowledge. Although some of the findings in the study match simple intuition of user behaviors, by extensive investigation we were able to further quantify and show the minor differences in the details systematically and reason about the cause of those differences (e.g., user population, network environment, time of trace collection, etc.). Second, by proposing metrics for describing individual MN behaviors, we propose a basis on which mobility models for individual MNs can be established. Third, we identify several facts (e.g., skewed location visiting preferences, low unique encounter ratio, etc.) indicating that conventional synthetic mobility models (such as random waypoint, random walk, etc.) are not adequate for a heterogeneous environment such as corporations and university campuses.

It is thus very important to *realistically* understand the underlying environments, based on a thorough analysis of empirical data sets. Such an understanding sheds light on fundamental structure in user behaviors, and one should drive user models and protocol design with these observations.

References

[1] MobiLib: Community-wide Library of Mobility and Wireless Networks Measurements. http://nile.usc.edu/MobiLib
[2] CRAWDAD: A Community Resource for Archiving Wireless Data at Dartmouth. http://crawdad.cs.dartmouth.edu/index.php

[3] C. Perkins. *Ad hoc networking*, Addison-Wesley, 2000.

[4] W. Hsu and A. Helmy. On important aspects of modeling user associations in wireless LAN traces. In *Second International Workshop on Wireless Network Measurement (WiNMee 2006)*, April 2006, pp. 1–9.

[5] W. Hsu and A. Helmy. On nodal encounter patterns in wireless LAN traces. In *Second International Workshop On Wireless Network Measurement (WiNMee 2006)*, April 2006, pp. 1–10.

[6] W. Hsu, T. Spyropoulos, K. Psounis, A. Helmy, Modeling spatial and temporal dependencies of user mobility in wireless mobile networks, *IEEE/ACM Trans. on Networking*, vol. 17, issue 5, pp. 1564–1577, Oct. 2009.

[7] W. Hsu. Behavioral analysis, user modeling, and protocol design based on large-scale wireless network traces. Ph.D. dissertation at University of Florida, August 2008. Available at http://nile.cise.ufl.edu/˜weijenhs/.

[8] W. Hsu, D. Dutta, and A. Helmy. CSI: A paradigm for behavior-oriented delivery services in mobile human networks. Technical report available at http://arxiv.org/abs/0807.1153.

[9] M. Balazinska and P. Castro. Characterizing mobility and network usage in a corporate wireless local-area network. In *Proceedings of MobiSys 2003*, pp. 303–316, May 2003.

[10] M. McNett and G. Voelker, Access and mobility of wireless PDA users, *ACM SIGMOBILE Mobile Computing and Communications Review*, vol. 9, issue 2, pp. 40–55, April 2005.

[11] T. Henderson, D. Kotz, and I. Abyzov. The changing usage of a mature campus-wide wireless network. In *Proceedings of ACM MobiCom 2004*, September 2004, pp. 187–201.

[12] D. J. Watts and S. H. Strogatz. Collective dynamics of "small-world" networks. *Nature* 393:440–442, 1998.

[13] F. Bai and A. Helmy. A survey of mobility modeling and analysis in wireless ad hoc networks. In *Wireless Ad Hoc and Sensor Networks*, Springer, 2006.

[14] F. Bai and A. Helmy. Impact of mobility on last encounter routing protocols. In *Proceedings of the Fourth Annual IEEE Communications Society Conference on Sensor, Mesh and Ad Hoc Communications and Networks (SECON 2007)*, June 2007, pp. 461–470.

[15] F. Bai, N. Sadagopan, and A. Helmy. IMPORTANT: A framework to systematically analyze the Impact of Mobility on Performance of RouTing protocols for Adhoc NeTworks. In *Proceedings of IEEE INFOCOM 2003*, April 2003, pp. 825–835.

[16] R. Hogg and E. Tanis. *Probability and Statistical Inference*, 6th ed., Prentice Hall, 2001.

[17] R. Albert and A. Barabasi. Statistical mechanics of complex networks. *Review of Modern Physics*, 74(1): 47–97, January 2002.

[18] A. Vahdat and D. Becker. Epidemic routing for partially connected ad hoc networks. Technical Report CS-200006, Duke University, April 2000.

[19] S. Tanachaiwiwat and A. Helmy. On the performance evaluation of encounter-based worm interactions based on node characteristics. In *ACM Mobicom 2007 Workshop on Challenged Networks (CHANTS 2007)*, Montreal, Quebec, Canada, September 2007, pp. 67–74.

[20] W. Hsu and A. Helmy. MobiLib USC WLAN trace data set. Available from http://nile.cise.ufl.edu/MobiLib/USC_trace/

[21] D. Kotz, T. Henderson, and I. Abyzov. CRAWDAD data set dartmouth/campus/movement/01_04 (v. 2005-03-08). Available from http://crawdad.cs.dartmouth.edu/dartmouth/campus/movement/01_04.

[22] M. Balazinska and P. Castro. CRAWDAD data set ibm/watson (v. 2003-02-19). Available from http://crawdad.cs.dartmouth.edu/ibm/watson.

[23] M. McNett and G. M. Voelker. Wireless Topology Discovery project data set. Available from http://sysnet.ucsd.edu/wtd/

[24] J. Scott, R. Gass, J. Crowcroft, P. Hui, C. Diot, and A. Chaintreau. CRAWDAD trace cambridge/haggle/imote/infocom (v. 2006-01-31). Available from http://crawdad.cs.dartmouth.edu/cambridge/haggle/imote/infocom.

[25] D. Kotz, T. Henderson, and I. Abyzov. CRAWDAD trace set dartmouth/campus/tcpdump (v. 2004-11-09). We use the list of device types in the fall03 tcpdump data. Available from http://crawdad.cs.dartmouth.edu/dartmouth/campus/tcpdump.

Chapter 7

Quality of Service in an Opportunistic Capability Utilization Network

Leszek Lilien
Western Michigan University and Purdue University

Zill-E-Huma Kamal
Western Michigan University

Ajay Gupta
Western Michigan University

Isaac Woungang
Ryerson University

Elvira Bonilla Tamez
Ryerson University

Contents

7.1 Introduction

Pervasive computing is a powerful paradigm that indicates the goals for research and development of integrated computing and communication systems. It should be noted that *ubiquitous computing*, though often interpreted as a synonym for pervasive computing, should rather be viewed as its extreme and most advanced realization. The paradigm has the potential to provide extremely user-friendly systems that are "pervasively and unobtrusively embedded in the environment, completely connected, intuitive, effortlessly portable, and constantly available" [62].

Pervasive computing is a vision—a very broad and high-level paradigm. It can be realized in many ways, via different lower-level paradigms within its realm. We discuss quality-of-service issues for Oppnets, a new lower-level pervasive computing paradigm that we have proposed [1,2,3,4,6]. *Oppnets* foster the opportunistic growth of a network through the integration of available pervasive resources and their capabilities. We will introduce Oppnets with a use scenario in the next subsection, followed (in Subsection 7.1.3) by a more detailed description of their operations and characteristics.

7.1.1 Oppnet Use Scenarios

As an introduction to Oppnets and an indication of their potential, we consider an everyday situation.

Suppose that you are driving home from your workplace and receive a text message (read to you via a voice interpreter) from your spouse reminding you to go grocery shopping. You drive to the local grocery, but before entering the store you realize that you do not have the shopping list. You try to call your spouse at home but your phone fails, giving you the "No Signal" error message. (Should the crowd of callers at a nearby football game be blamed?) You head home frustrated by the waste of money, time, and energy.

A system implementing the Oppnet paradigm could have prevented your aggravation. Let us look at how the above scenario changes if you could use Oppnets. In Oppnets, an initial network, referred to as the *seed Oppnet* (or just the *seed*), receives jobs, such as "get the shopping list from home." In our scenario, your cell phone is the seed (it is the extreme case of a single-device seed Oppnet). Though connectivity via the cellular network is not working, there are other communication media that the seed can utilize. For example, most cell phones are Bluetooth-enabled. The seed can exploit this medium to search for other Bluetooth devices within its range. Once it has detected a Bluetooth device, say a PDA, it can find out what capabilities (resources, services, skills, etc.) it possesses. If the PDA belongs to a cellular network, the seed can negotiate a cell phone call over this Bluetooth connection to the PDA, and through it to the cellular network. (This is similar to the Bluetooth setup in cars for hands-free calling.) Now you can talk to your spouse to find out what to buy.

What if the seed was unable to find any devices that were connected to a cellular network via Bluetooth, Wi-Fi, WiMAX, wired Internet, or any other communication network? (Note that the seed-to-device connection can be multihop rather than single-hop, as was the case in the example above. Furthermore, each hop can use an arbitrary communication technology to contact the next device in the chain from the seed to the cellular network.) In this case, another Oppnet scenario preventing your frustration is possible.

Suppose that indeed the seed was unable to find in its range any devices that were connected (directly or not) to a cellular network. Suppose further that you had set up a *home area network* (*HAN*), and that it includes your smart refrigerator (e.g., LG Electronics Internet Refrigerator [61], which can provide the grocery shopping list to any authorized Internet-connected device or user. Suppose that now the seed, running on your cell phone, does not find a PDA within its range, but only a Bluetooth-enabled MP3 player, which does not offer a direct connection to the

Internet. The seed can negotiate with the MP3 player to enlist its help in searching for other devices that could help in searching for Internet services or offer Internet services themselves. (Importantly, these will be devices that are beyond the range of the seed, so the seed cannot reach them directly.) For example, the MP3 player could reach another cell phone, which can in turn reach a laptop that has a Wi-Fi Internet connection.

The laptop will be the final link in the chain of devices passing your information request to the fridge on your HAN and obtaining the shopping list from the fridge. The shopping list is then sent back through the Internet and Bluetooth devices to the seed and displayed on your cell phone. Again, you can complete your shopping without loss of time, money, and other resources.

These scenarios show the intrinsic property of Oppnets—they search for disparate devices or networks that can serve as *helpers* whenever needed to accomplish a job. Helpers are then integrated into the seed Oppnet—despite heterogeneity in hardware, software, protocols, communication infrastructure, etc.—to utilize available capabilities that exist in the ambient environment of the seed (*capabilities* include resources, services, skills, etc.). In the process, the seed Oppnet grows into a larger network, known as an *expanded Oppnet*. This growth of the seed into an expanded Oppnet is a fundamental property of Oppnets and also their salient feature, distinguishing them from other prevailing and proposed technologies.

7.1.2 Classes of Opportunistic Networks and Oppnets

Opportunistic networks known in the literature can be roughly categorized into three classes. *Class 1 Opportunistic Networks*—or, more descriptively, *Opportunistic Communications Networks*—are most restricted in the usage of capabilities available in their environment. Their only goal is to establish communication capabilities by interconnecting previously disconnected devices. In this class, devices or networks that are within the communication range of each other initiate connection and dynamically configure an ad hoc network, or a store-and-forward network, when there is a need for such communication. Class 1 Opportunistic Networks are discussed and studied in [8,63,64,65,66,67,68], for example.

Class 1.5 Opportunistic Networks—or, more descriptively, *Opportunistic Data Dissemination Networks*—have a broader goal than Class 1; that is, to opportunistically disseminate data. Here, devices or networks initiate a dynamic interconnection with other detected components (devices or networks), with the purpose of propagating data. We call it *Class 1.5* because it might not deserve an integer-numbered category, being a relatively simple extension of Class 1. Examples of Class 1.5 Opportunistic Networks are shown in [7,9,43].

Class 2 Opportunistic Networks—or, more descriptively, *Opportunistic Capability Utilization Networks*—aim at using all sorts of capabilities. Their capabilities include resources, services, skills, and so on, such as processing power, storage, sensing and actuating, and specialized skills (including the ones provided by software).

In particular, capabilities are not limited to those involving communication or data dissemination. Comprehensive proposals for paradigms or technologies that can be classified as Class 2 Opportunistic Networks are limited (e.g., [28]).

The salient features of Oppnets, distinguishing them from other conceivable realizations of Class 2 Opportunistic Networks, are as follows. An Oppnet grows from an initial seed into an expanded Oppnet by integrating helpers (they would, otherwise, remain disjointed devices or networks). The initial and the new components of an expanded Oppnet work together to perform Oppnet's tasks. This shows that growth of an Oppnet is an inherent side effect of this computing paradigm. An expanded Oppnet can shrink when helpers that are no longer needed are released by it. This specific mode of growth/shrinking capability is a distinguishing factor when compared to other prevailing or proposed pervasive and opportunistic computing technologies.

Class 2 Opportunistic Networks were first mentioned as Opportunistic Sensor Networks in Bhargava et al. [1]. Opportunistic Sensor Networks were later generalized to Oppnets [2]. In subsequent publications, we explored research challenges and security and privacy issues for Oppnets [2,3]; defined a standard implementation framework for Oppnet applications with an application programming interface, or API (Oppnet Virtual Machine or OVM) [10,11]; designed and implemented a small-scale proof of concept for Oppnets [11,12]; and presented a comparison of Oppnets with other ad hoc networking technologies specialized for emergency applications [5].

7.1.3 Basic Oppnet Operations and Categories of Helpers

Suppose that a seed Oppnet is deployed within a certain existing pervasive computing environment or infrastructure. (As in any other pervasive computing technology, they will work better and better within a more and more pervasive infrastructure—that is, with more capabilities available in the surrounding pervasive computing milieu.) The seed can comprise a network of wireless, mobile, ad hoc, and other powerful computing components. It can also include "thin"—that is, less-capable, resource-constrained—embedded devices.

After the seed self-configures and becomes operational, its first challenge is to analyze the jobs at hand and the capabilities necessary to complete them. If there is a deficiency in capabilities, the seed forages through its cyber environment to find the needed capabilities.

At any given moment, components that belong to an Oppnet can be called its *insiders*, and the ones that do not belong to the Oppnet, *outsiders*. All outsiders are candidates for Oppnet helpers, depending only on the Oppnet's needs and whether they can be reached by the Oppnet (via a chain of helpers).

We can distinguish two main categories of Oppnet outsiders. The first category, *Oppnet-enabled* outsiders, are those that are easy for an Oppnet to communicate with and get help from because they speak the Oppnet's language: They contain primitives and protocols for communication as well as requesting and providing help. For example, they contain primitives and protocols defined by the Oppnet Virtual Machine (OVM) [10,11]. The second category, *Oppnet-unenabled* outsiders, is the remaining ones, which lack such facilities. As a consequence, Oppnets contacting them must expend much more effort, including learning to speak their language, when attempting to contact them, and then use them as their helpers.

Before using any outsider as a helper, an Oppnet must first obtain the outsider's agreement to participate. Many incentives might be used by the Oppnet to convince a contacted outsider to become its helper (including economic ones, such as micropayments) [42,24].

We defined a special subcategory of Oppnet-enabled outsiders, known as *Oppnet reservists*, which have agreed a priori to help an Oppnet asking for their help. (This is analogous to, e.g., Air Force reservists who have agreed a priori to "help" the Air Force any time they are "asked" to do so.) Incentives might again be involved in the decision of any device or network to sign up as an Oppnet reservist. (This is analogous to Air Force reservists signing up for incentives ranging from intangible ones, like patriotism, to material ones, like receiving tuition and stipends.) A given reservist could specify, in the Oppnet Reserve Agreement, his willingness to help only a special application category of Oppnets. For example, a reservist could agree to help at any time if the call for help comes from an Emergency Response Oppnet but can deny requests coming from, for example, a City Maintenance Oppnet. (Although weak, this would be analogous to an Air Force reservist agreeing to help only the U.S. Air Force, but not the U.S. Navy or the Polish Air Force.)

Before being able to contact any helper candidate, the Oppnet must first discover it. The discovery process could be as simple as a capability directory lookup in search of listed outsiders (which is a pseudo-discovery) or as complex as an ad hoc *true discovery* of unlisted outsiders. For a given category of Oppnets, their reservists should be listed in their private directory, which will significantly simplify and speed up the discovery process. For instance, Emergency Response Oppnets would have the directory of Emergency Response Reservists.

An Oppnet evaluates the helper candidates and invites desirable candidates to join it—these are helpers that can offer missing capabilities. In general, each nonreservist candidate is free to accept or to reject the invitation. However, under certain circumstances (e.g., in life-or-death emergencies) even nonreservist candidates can be ordered to join the Oppnet [2,3]. Once a candidate accepts an invitation to join the Oppnet as a helper, its capabilities (or their subset agreed upon by the Oppnet and the candidate) are integrated with the resources of the Oppnet. In this way, if the Oppnet is a seed Oppnet, it grows into an expanded Oppnet; otherwise, an expanded Oppnet grows into a larger expanded Oppnet.

This iterative process is repeated every time more resources or capabilities are needed, so the Oppnet grows into a larger and more powerful system spontaneously, on an as-needed basis. It is important to note that Oppnets are not restricted from growth due to hardware, software, protocol, primitives, or communication media discrepancies. In this manner, Oppnets not only provide opportunistic use of diverse capabilities, but also grow by integrating capabilities of disjoint devices or networks (outsiders). Oppnets can also shrink—even back to the seed Oppnet size—by releasing helpers who have completed their tasks (or failed in executing the expected tasks).

Numerous applications for Oppnets can be envisioned, from the everyday (similar to the opening scenario) to more intricate ones, and from nonsensitive applications to mission-critical military applications. Examples are presented by Lilien et al. [3].

7.1.4 Quality of Service in Oppnets

Different definitions of quality of service (QoS) depend on the context in which the concept of quality is applied. However, they all have something in common: the definition of an expected behavior. The International Telecommunications Union (ITU) defines QoS as: "a collective effect of service performances that determine the degree of satisfaction by a user of a service" [38]. The International Standard Organization (ISO) defines QoS as: "a set of qualities related to the collective behavior of one or more objects" [39]. Anbazhagan and Nagerajan [40] state that "QoS determines the service usability and utility which influence the popularity of the service."

Some of the common *QoS requirements* for Web services are [40]: (i) *availability*—indicating if the service is ready for use; (ii) *accessibility*—indicating the likelihood of a successful instantiation of a service; (iii) *integrity*—indicating the correctness of the interaction with other Web services; (iv) *performance*—indicating throughput, and latency or delay; and (v) *reliability*—indicating the ability of maintaining service quality (confidentiality, access control, message encryption, etc.). We consider only the performance QoS requirements for Oppnets: throughput and delay, which measure effectiveness and efficiency.

It is often necessary to assure that certain QoS requirements are met in a computing system in order to ensure that jobs are effectively and efficiently completed. This is even more challenging in Oppnets, when the initial Oppnet, the seed, does not possess all the capabilities it needs to complete its jobs (or complete the job faster or better—with these extra capabilities).

Many QoS parameters are application-dependent, which complicates the QoS issue. For example, in military applications, real-time response (with hard deadlines) is imperative. In any applications that require meeting soft or hard deadlines, users must be allowed to define QoS requirements (including throughput and delay). An Oppnet realizing users' jobs needs to invoke its own and helpers' capabilities and utilize them in such a way that the QoS demands are met.

The need for a mechanism to automatically interpret available Web content without human intervention has led to the development of a new vision for the next Web generation, known as the *Semantic Web* [31]. In this new paradigm, online services and software agents, assisted by *ontologies* (describing the application domain, determining its vocabulary, and describing the relationships of its elements), interact in an autonomous way to satisfy users' requirements. Services are described by service specifications. A user that needs a particular service can find the needed service. Service specifications are not limited to service *functions* or characteristics, but include *nonfunctional* properties, such as QoS, which define requirements.

In Oppnets, the QoS requirements describe the behavior of a capability, as expected by a human or artificial user. QoS requirements are also used to further refine the *capability discovery* and the subsequent *helper-matchmaking process*. It matches user needs with the descriptions of capabilities of helper candidates to find helpers—potential capability providers.

One of the contributions of this chapter is introduction of a novel solution for implementing Semantic Web capabilities and QoS requirement specifications for Oppnets.

7.1.5 Chapter Organization

Section 7.2 presents a resource utilization technique called *service location and planning* (*SLP*)—which uses QoS parameters for resource utilization, illustrates it with a small-scale Oppnet scenario, compares SLP to other mathematical models and optimization problems, and discusses related work. Section 7.3 presents background material and work related to the Semantic Web, semantic service discovery, using Semantic Web for QoS requirement specification, and the Oppnet SCOW-Q model for QoS requirements in Oppnets. Section 7.4 concludes with a brief summary of the chapter.

7.2 QoS and Resource Utilization in Oppnets

Resource utilization in Oppnets is similar to use of services in service-oriented networking. Pervasive service-oriented computing technologies often abstract resources to services, where services can be e-mail applications, stock quotes, weather updates, communication channels, etc. We use the terms *services* and *resources* interchangeably, and they are the only kinds of *capabilities* we discuss in this section.

In an Oppnet, resources are scattered across the network and may be offered to an entity requesting them by nodes possessing different processing power and communication bandwidths. For optimal efficiency and effectiveness, the node that can provide the fastest access to a needed resource should be chosen to satisfy the request.

Two factors, *latency* (or *delay*) and *throughput*, are the most fundamental measures of the quality of service for consumers. Other QoS measures are reliability, efficiency, scalability, accuracy, and so on. However, before these QoS parameters can be studied, it is imperative to provide the necessary (e.g., at least the requested) throughput with the minimum acceptable delay.

In this section, we discuss methods for identifying the resources in an Oppnet and a service location and planning (SLP) mechanism for optimizing resource utilization in Oppnets.

7.2.1 Service Location and Planning Mechanism for Resource Utilization in Oppnets

The SLP problem [29,30] can be defined with the following set of givens:

1. A network graph $G = (V, E)$. V is a set of vertices representing computing nodes. E is a set of edges representing established communication links between the nodes.
2. A set of services S offered by the producers in the network represented by G.
3. A set of consumers, $V^C \subseteq V$, such that each node $v \in V^C$, representing a consumer, is requesting a service $s_i \in S$, $i = 1,..., |S|$. Each service request has an associated throughput and delay parameter.

There is a cost associated with using the service s available at the node n in G. This cost may include: (i) the cost of accessing node n, expressed as the number of hops between the consumer and node n, (ii) the cost of using storage at node n, (iii) the cost of using processing resources of node n, (iv) if necessary, the (appropriate part of the) cost of updating node n so that it can use the latest version of the servicing application, and (v) the costs of using incoming and outgoing communication links of node n. When multiple services are used by a consumer from the same node n, a discount Γ is given to the consumer. This approach is used to promote service federations at service nodes (i.e., using multiple services from the same node).

Using multiple services from the same node can be beneficial in many ways. For instance: (1) it enables aggregation and compression of data for one customer, enabling better bandwidth utilization; (2) the cost of accessing a node for service invocation can be spread across many service requests of a consumer; (3) security- and privacy-related negotiations are made by a consumer with one node, which reduces costs for trust management and enables faster service invocation.

The problem is defined as follows: Use services from a set of producers (providers), $V^P \subseteq V$, in such a way that the incurred service invocation costs are minimized, while the QoS demands of throughput and delay are satisfied, and the underlying network link capacities are not exceeded. (Producers can also be consumers of services, which implies $V^P \cap V^C \neq \varnothing$.)

7.2.2 Using SLP Model for Resource Utilization in Oppnets

The above SLP problem has been modeled as an integer linear programming (ILP) problem and has been solved optimally for both small-scale and large-scale networks using two techniques: linear programming solve engines [29] and Lagrangean relaxation [30]. No details are given here due to space limitations (we encourage the interested readers to see the two referenced papers). We focus here exclusively on using the SLP model for resource utilization in Oppnets.

7.2.3 Scenarios Illustrating Use of SLP in Oppnets

Let us consider a very simple illustrative example showing use of SLP in a small-scale Oppnet. A house is equipped with a laptop (with a built-in webcam), smoke detectors, and security webcams as illustrated in Figure 7.1. Suppose that the laptop is the seed of the home Oppnet and that all smoke detectors were invited to join the Oppnet, became its helpers, and are connected to the Oppnet via Bluetooth links. Suppose that the security webcams are Oppnet-enabled but were not invited to join the Oppnet.

The laptop receives data on smoke concentration levels from the detectors, which are able to detect miniscule traces of smoke well before a fire starts. The laptop intelligently interprets the collected sensor information, deciding when the *smoke threshold* is exceeded and the owner should be notified. When this happens, the laptop can notify the owner via her cell phone. She can ask the laptop for a video feed from the endangered area of her house. The laptop finds that the security webcams are candidate helpers capable of providing the feed (its built-in webcam is not as good as the external webcams). It invites one or more security webcams from the neighborhood of the smoke detector that sent the alarm. The invited webcams become Oppnet helpers, connected to it via Wi-Fi. Once the security webcams are integrated into the Oppnet, the laptop passes control over them to the owner. She can now direct a security webcam to turn and focus on any area within its range (in the near future, the webcams can become independently mobile, or robot mounted). By a visual evaluation of the situation, the owner may be able to eliminate false alarms. For example, she might notice that the smoke comes from a cigarette belonging to an inconsiderate new cleaning person. (In the future, an intelligent image recognition agent will be able to focus the webcam on a source of smoke automatically.)

Figure 7.1 A home equipped with Oppnet-enabled devices.

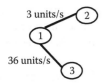

Node	Service A (feed video)	Service B (sense smoke)	Service C (commu- nicate)	Service D (process)
1 (laptop)	100	5,000	10	10
2 (smoke det.)	5,000	5	5,000	5,000
3 (webcam)	5	5,000	5,000	5,000

(a) Input 1 for the SLP problem: The Oppnet topology for the experiment, incl. the laptop (1), the smoke detector (2), and the security webcam (3).

(b) Input 2 for the SLP problem: Service costs for services A, B, C and D in the given topology.

Node	Throughput Requirements		Delay Requirements	
	Service A	Service B	Service A	Service B
1 (laptop)	20	2	10 s	10 s
2 (smoke det.)	0	0	0 s	0 s
3 (webcam)	0	0	0 s	0 s

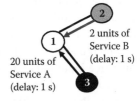

2 units of
Service B
(delay: 1 s)

20 units of
Service A
(delay: 1 s)

(c) Input 3 for the SLP problem: QoS requirements for service requests.

(d) Output: The solution for the SLP problem.

Figure 7.2   Resource utilization in Oppnets using the Service Location and Planning (SLP) technique.

The SLP technique can help realize this scenario in an efficient way. For simplicity, suppose that only one smoke detector and one security webcam are used.

Node 1 (i.e., the Oppnet seed implemented by the laptop) detects and integrates as Oppnet helpers Node 2 (the smoke detector) and—when requested for a video feed—Node 3 (the webcam). Figure 7.2a shows all three nodes, their links, and link speeds. The indicated data rates of 3 units/s (Mb/s) and 36 units/s are typical for Bluetooth and Internet links connecting Node 2 and Node 3, respectively, with Node 1. The Oppnet of Figure 7.2a includes, as depicted in Figure 7.2b, four services: feeding video, sensing smoke concentration levels, providing communication facilities, and providing processing power, denoted as Services A, B, C, and D, respectively.

Node 1 asks Nodes 2 and 3 about the costs of services they can provide. It records the costs of these services, as well as the costs of its own services, in the table presented in Figure 7.2b. The cost of 5000, shown in the table, is sufficiently large to be used as an infinite cost, indicating that the node represented by the given table row cannot provide the service identified by the given table column. For instance, a cost of 5000 in the cell located in the row for Node 1 and the column for Service B indicates that Node 1 cannot provide Service B (the laptop has no built-in sensor to provide smoke detection). In some cases, a node looking for a helper has only one candidate helper that is able to provide a needed service. For instance, Service B can be provided only by Node 2; the costs of getting Service B from Node 1 or Node 3 are infinite (5000). In other cases, a node looking for a helper finds multiple candidate helpers able to provide a needed service. For instance, Service A (video feed) can be provided by Node 1 (the laptop's built-in webcam) at the cost of 100 and by Node 2 (the security webcam) at the cost of 5; of course, the latter helper, offering a lower-cost service, is more attractive.

Figure 7.2c expresses the service requests of all three nodes. It indicates that Node 1 needs 20 units of Service A (a video feed) and 2 units of Service B (smoke sensing), both with no more than 10 seconds of delay. Nodes 2 and 3 request no services.

Three input data sets are used by the ILP model. Figure 7.2a shows the Oppnet topology, including bandwidth restrictions. From Figure 7.2b, ILP knows that the only way the Oppnet can get smoke concentration levels is via Service B of Node 2 (the smoke detector) and that the optimal (but not only) way the Oppnet can get a video feed is via Service A of Node 3 (the security webcam). From Figure 7.2c, ILP knows which nodes to ask for services and the QoS requirements for the services.

The three input data sets are used by the ILP model to yield the solution presented in Figure 7.2d. It shows that the optimal resource utilization for the Oppnet in this scenario is to use 20 units of Service A from Node 3 and 2 units of Service B from Node 2. This solution meets the throughput and delay requirement as follows: 20 units of Service A can be accommodated by the 36-unit link, and 2 units of Service B can be accommodated by the 3-unit link—with the delay of 1 s for each service delivery. This satisfies their delay requirements of 10 s (shown in Figure 7.2c). In this way, service is used while abiding by consumer-defined QoS parameters and the underlying link-layer bandwidth constraints.

The illustrative purposes demanded use of a very small example. In larger and more complex Oppnet configurations and scenarios, we might need to obtain portions of requested services from multiple nodes to meet delay requirements. For example, if we needed a video feed of 361 Mb, receiving the video over the 36-Mb/s link would cause a delay of 11 s, violating the QoS delay constraint. We could meet the constraint by finding other helpers—e.g., the laptop's built-in webcam—so that portions of a video could be provided by multiple service providers in parallel, satisfying the service request. We could meet the QoS constraint with this type of service splitting. Examples of solutions of the SLP problem for large-scale networks are provided elsewhere [29,30].

7.2.4 Related Work on Optimization Problems and Pervasive Computing

This section looks at various optimization problems and pervasive computing technologies to show the novelty of the SLP model.

7.2.4.1 Related Work on Optimization Problems

The *capacitated facility location* (CFL) problem [21] can be stated as follows. Given a set of facilities F and a set of clients C, each $c \in C$ has a demand d_c that must be serviced by one or more open facilities. There is a facility cost f_i for opening facility $i \in F$ and a service cost $s_{i,c}$ for facility i to service one unit of demand from client c. No facility may service more than U units of demand. The goal is to service all clients at the minimum total cost (i.e., the sum of facility and service costs). The CFL problem is NP-hard (non-deterministic polynomial-time hard) even in the case where U = ∞, which is known as the *uncapacitated facility location* (UCFL) problem [13,14,21].

Compared to CFL, the SLP problem has an additional dimension, namely the QoS constraints to meet delay requirements. This makes the SLP problem more complex than the CFL or UCFL problems.

Other problems of interest—such as the *optimal placement of gateways in wireless mesh networks* [15,16] and *service selection minimizing costs or maximizing QoS* [17] problems—are also considered variants of the facility location problems.

Related optimization problems are known in the area of *graph theory*. It may seem that the vertex cover problem and its variants are similar to the SLP problem. However, a closer look at vertex cover problems reveals that they do not allow for formulation of QoS constraints. Vertex cover problems in undirected graphs [18] are different, since SLP uses directed graphs. Furthermore, these problems cannot be used to guarantee optimal solutions for the SLP problem, whereas the ILP model of the SLP problem gives optimal solutions for the SLP problem. However, directed graph vertex cover problems [19] could be used as a basis for heuristic and meta-heuristic programming for the SLP problem.

A review of *networking-related optimization problems* shows that problems pertaining to service or content replication are similar to the SLP problem. Replication techniques of Jin and Wang [22] can be used in our future work to analyze the trade-off between the benefits of service replication and costs of additional service installation.

Another alternative to service replication is *service composition* as studied by Gao et al. [23]. We account for service composition in the SLP problem through service federation constraints.

7.2.4.2 Related Work on Pervasive Computing Technologies

Technologies such as Service Discovery Protocols (SDP), known from the *pervasive computing* and *service-oriented computing* paradigms, can be studied to gain insight for resource utilization and invocation. Zhu et al. [20] present a comparison of existing protocols that implement service discovery and communication—for instance, Jini, UPnP, and Bluetooth SDPs. A disadvantage of these older SDPs is the need for preconfiguration of the participating devices with common protocols, which inhibits creating truly pervasive or open service-oriented computing environments. This deficiency is avoided in newer SDPs used in more recent networking technologies—such as Oppnets, AONs, and so on. Oppnet operation is more efficient when facilitated by installation of Oppnet primitives (such as OVM [10,11]), but an Oppnet is able to function even without such preconfiguration.

Summarizing, the inadequacy of the known optimization and graph-theoretic problems as well as service discovery protocols motivated our research on the SLP problem. SLP not only helps consumers who demand services with QoS parameters, but also minimizes costs for consumers while operating within the bandwidth capacities of the underlying network.

The SLP technique can be used not only in Oppnets. It is applicable to the new technology of Application-Oriented Networking (AON) in service-oriented computing, launched by a major player in the field of communications technology [25].

7.3 Using Semantic Web to Specify QoS Requirements in Oppnets

Most information on the Web was designed to be read and understood by humans. Computer programs have the ability to interpret only some attributes of this content. However, there is a limitation in understanding and processing the meaning of Web contents. To address this deficiency, a vision known as the *Semantic Web* was introduced [31]. The goal of the Semantic Web is to foster an environment where *software agents* (intelligent programs, a.k.a. *agents*) perform sophisticated tasks on behalf of users, automatically processing information with a well-defined meaning, resulting in a successful collaborative effort among humans and computers.

7.3.1 Basics of the Semantic Web

The Semantic Web uses two technologies that complement each other: the *eXtensible Markup Language* (*XML*), and the *Resource Description Framework* (RDF). XML allows building a structure for contents and documents but does not provide a clear meaning of the structure. RDF complements XML by providing this meaning.

Services are bundles of functionality that are advertised to be used or requested. They play a key role in the Semantic Web because they are responsible for collecting the content, processing it, and exchanging it. They can even allow agents that were not designed to work together to collaborate on data transfers when determined so by the semantics.

Service descriptions include information on the service's interface and metadata that can be used during the service discovery process, ideally all in a machine-understandable language. The term *profile* refers to documents that describe the capabilities of a service or a device [34].

Services are typically identified by a *Unified Resource Identifier* (URI) and rely on technologies for message exchange (such as XML) and transport protocols (such as HTTP and SOAP, Service-Oriented Architecture Protocol). A URI can be used to identify physical resources, such as cell phones, TV sets, and so on. The RDF language can describe these resources and the way in which they can operate together as if they were agents [31].

The Semantic Web is challenged in its capacity to express data and rules for reasoning. It faces a problem where one meaning is defined by two or more different terms. This issue can be addressed by including a collection of *ontologies* (i.e., structured vocabularies) that define the relationships among *terms* or *concepts* (also called *classes*). Ontologies enhance the communication and identification of services that are available to satisfy a given service request. For instance, the use of ontologies in profiles enables automatic search for and discovery of needed functionality in devices and employment of well-described services.

The ontology for the Web is formed by a taxonomy and a set of inference rules [31]. The *taxonomy* defines classes of objects and their relationships. *Inference rules* provide the processing mechanisms to manipulate terms in a meaningful way. Ontologies are commonly expressed by *Ontology Web Language* (OWL), which has its foundation in description logics (DL) [32]. OWL allows for inferring of relationships between concepts and their definitions. It is an evolution of its predecessors: *DARPA Agent Markup Language* (DAML) and *Ontology Inference Layer* (OIL).

There are alternatives for enabling the semantic description of ontologies. The W3C organization developed a framework known as the *Web service architecture* (WSA) [33] to provide standard definitions for the Web services model and the relationships among its different components. *Web Services* (WSs) have been defined as "a software system designed to support interoperable machine-to-machine interaction over a network" [33]. WSs are characterized by two main entities—the providers and the requesters of services. These entities exchange messages through the use of a *Web Service Description Language* (WSDL), which defines the functional specifications of services (for instance, their data types, formats, and protocols). *Semantics* in WSs refers to the expected behavior during service interactions. It represents a contract or an agreement between the participants.

7.3.2 Semantic Web Services and Their Specification

Web services and the Semantic Web complement each other, evolving into what is known as *Semantic Web Services* (SWSs) [59]. SWSs are self-contained and self-described entities that semantically advertise their capabilities and descriptions and assist in the discovery, composition, and binding of services. SWSs use machine-understandable language supported by ontologies and function in an open, heterogeneous environment.

Service discovery is usually performed by agents that look for service descriptions matching the desired functional or semantic criteria specified in the user's service request. There are two important aspects of service discovery: the *architecture*, determining where services can be broadcast or advertised [37], and the *matchmaking mechanism*, determining how services are compared against the user's requests until a proper match is found. Typically, only the *functional* characteristics of a service are used in service matching, ignoring the *nonfunctional* properties, such as QoS requirements. Nonfunctional properties add extra service-matching criteria, thus narrowing the number of functionally equivalent Web services.

7.3.2.1 Semantic Web Service Specification

Approaches facilitating a semantic specification of Web services include the following:

- The DARPA program [59] proposed DARPA Agent Markup Language (DAML), which extends XML and RDF. The Ontology Inference Layer (OIL) adds the underlying ontology layer to the Semantic Web. DAML and OIL were succeeded by DAML+OIL, which includes features of both its predecessors, and evolved into the Ontology Web Language (OWL).
- The DAML organization proposed the use of OWL as the "representation language of choice for the OWL-S proposal." (OWL-S was formerly known as DAML-S.) The ontology of services proposed by the OWL Service Coalition [35] provides information on: (a) the *service profile*, which presents each service and describes its functionality and characteristics; (b) the *service model*, which describes how the service works; and (c) the *service grounding*, which provides details on how a service can be accessed.
- The *Web Services Modeling Ontology* (WSMO) [60] allows for service specifications and provides support for addressing inaccuracies in describing services. It uses the *Web Service Modeling Language* (WSML).
- In an effort to incorporate semantics into descriptions of the Web services, W3C extended their WSDL language into the *WSDL-S*. WSDL-S is a standard format in XML, used to describe the "network services as a set of endpoints" [36] based on the syntax of the services. WSDL-S enhances the WSDL descriptions by adding semantics to them and by incorporating into them some concepts from OWL, WSMO, and so on. The benefits of WSDL-S over OWL-S are twofold: (1) semantic- and operation-level descriptions can be defined more clearly and (2) either UML or OWL can be used to represent ontologies.
- Recently, a standard called WSDL 2.0 has been introduced [37], providing "a model and an XML format for describing Web Services." It distinguishes between the *abstract description* of a service and the *concrete specification* of the offered functionality.
- *First-order Logic Ontology for Web Services* (FLOW), in addition to using XML and URIs for Web support, utilizes semantics defined by means of the first-order logic. FLOW incorporates standards such as WSMO, OWL-S, and process specification language (PSL) (ISO 18629). In addition, it supports a direct mapping to the *Rules Ontology for Web Services* (ROWS), a language based on logic-programming semantics from the DAML consortium.

7.3.2.2 Semantic QoS Specifications

Web services enable specification of both functional and nonfunctional service properties. The International Telecommunications Union (ITU) [38], the International Standard Organization (ISO) [39], and Anbazhagan and Nagerajan [40] indicate that QoS refers to nonfunctional properties of Web services. One might ask about the benefits of specifying nonfunctional requirements, especially QoS requirements, and how this influences service discovery.

Typically, nonfunctional properties have not been a part of the description of Web services. However, Tsesmetzis et al. [41] indicate that adding QoS features to profiles of Web services is beneficial, since the profiles are used in service discovery. They also proposed a QoS ontology language and a model to formally describe both QoS parameters and the relationships among these parameters. They not only focused on the QoS semantics (by developing vocabularies) but also developed a descriptive representation of QoS parameters for heterogeneous machines.

Some researchers incorporated QoS awareness not into service profiles, but into the Web services discovery mechanism. Tsesmetzis et al. [41] define the QoS ontology as a three-layer process, with each layer addressing different demands during the discovery process. For example, the QoS profile layer is used for service-matching purposes, and two more layers were proposed for property definition and metrics, respectively. No QoS ontology vocabulary was provided. Kim et al. [26] describe a framework called MOQ (Mid-level Ontologies for Quality), in which a set of ontologies defines the quality of service, and encompasses domain-independent concepts.

7.3.3 Related Work on the Semantic Web and QoS

In this overview of related work, we concentrate on the implementation of the SWS paradigm in different environments, as well as on the implementation of QoS, focusing on Web service discovery and service selection [44,69].

7.3.3.1 Semantic Web Services in Ubiquitous and Pervasive Computing

Although there is a tendency to consider ubiquitous computing and pervasive computing as synonyms, Gaber [45] distinguishes between the two based on the environment in which they operate and their interaction with it. *Pervasive computing* is considered to be adaptable and interacts with the closest environment, enhanced by context awareness and emergent functionalities. The purpose of *ubiquitous computing* is to provide users with a global access to services and devices anytime and anywhere. Gaber [45] provides the following classification of interaction paradigms: (1) the *client server paradigm* (CSP) is a traditional approach, in which a user places a request for a service already known to be available; (2) the *adaptive services to client paradigm* (SCP), suitable for ubiquitous computing, uses a decentralized and self-organizing agent-based approach to deliver a service to the user; and (3) the *spontaneous service emergence paradigm* (SSE) deals with unexpected and spontaneous creation of services, which are provided by nodes interacting in ad hoc connections, and is suitable for pervasive computing environments.

There have been other approaches to handle SWSs in other environments. For instance, for pervasive computing environments, Amigo-S [47] extends the OWL-S framework by integrating features that characterize the heterogeneity and richness of pervasive environments.

Due to the heterogeneous nature and user-centric goals of pervasive computing, the SWS approach enables ad hoc relationships between service providers and requesters using semantics. However, challenges such as the limitation of resources entering the network and the absence of a centralized mechanism to maintain registries and ontologies for service discovery make the implementation of this technology difficult. Ben Mokhtar et al. [58] indicate that significant computational effort is one of the major disadvantages of using semantic technologies for service discovery within pervasive environments. They present a competitive scheme called EASY (Efficient Semantic Service Discovery) for pervasive environments with QoS context and support. EASY is not yet another service discovery protocol, but it can instead be layered on top of existing ones, leveraging semantic abstractions at a higher level. This approach is composed of two parts: (1) the EASY Language (EASY-L) ontology used for semantic service description, which assures independence of underlying layers (or middleware infrastructure) and addresses the specification of nonfunctional properties such as QoS; and (2) the EASY Matching (EASY-M), which can be used to support matchmaking of nonfunctional services.

According to Chakraborty et al. [46], service discovery in pervasive computing should be decentralized, autonomous, self-advertised, and adaptable, in order to reflect environmental challenges. The authors presented a novel approach to service discovery in pervasive computing in which discovery architectures and service matchmaking tasks are coupled, building upon the concepts of peer-to-peer dynamic caching, service advertising, and a group-based forwarding of service discovery requests. The service description is supported by OWL.

Gagnes et al. [27] indicate that not only a richer description of services is needed in dynamic environments, but also better mechanisms to distribute them are required. Although *Universal Description, Discovery, and Integration (UDDI)* can be an instrument for providing Web service descriptions, in their opinion it is not appropriate for delivering semantic service advertisements in dynamic environments, where the term *dynamic* could be interpreted as a rapid and spontaneous change in topology or location, where information sharing occurs ad hoc. The authors proposed a generic service discovery architecture applicable to dynamic environments, that leverages concepts of Web services and a distributed multiregistry topology. They also present a categorization of service discovery topologies.

Ben Mokhtar et al. [47] presented COCOA, a solution for a conversation-based service composition with QoS support. COCOA consists of two mechanisms: (1) COCOA-SD for QoS-aware semantic service discovery, and (2) COCOA-SI for QoS-aware service integration. The authors also discuss the issue of *syntactic heterogeneity* of service descriptions in a pervasive environment, assuming that most agreements between service requesters and providers are based on a common service description syntax. To resolve this issue, they suggested implementing the semantic modeling of functional and nonfunctional service features through ontology-based semantic reasoning. COCOA-L is an OWL-S-based language for semantic service specification and semantic-aware description of services. This language is used by COCOA-SD to enable matching of service functionalities complemented by QoS-based matching.

7.3.3.2 Semantic Web Services in Mobile Ad Hoc Networks

Rapidly changing characteristics as well as the autonomous nature and decentralized topologies of MANETs make discovery of services a very challenging task. MANET dynamics prevent use of agreed upon or predefined service interfaces, while most of the existing models, architectures, and languages have been developed considering a universally connected environment, such as the one available in the Web. For this reason, we need to design a mechanism where an exchange of service representations can take place without using a formal representation. The use of SWSs is an alternative. However, it is necessary to have appropriate technologies that can handle the distribution of ontologies despite the spontaneous nature of MANETs.

Nedos et al. [48] presented a model for autonomous semantic service discovery assuming a role symmetry, which means that each node can potentially be both a service producer and a service consumer. This model differs from others because the semantic representation is not shared by the nodes, but is instead derived through node interactions. The authors also indicate that in order to apply standards for service discovery (such as WSMO, OWL-S, or WSDL-S) in a mobile ad hoc environment, the following requirements should be satisfied: (i) nodes should interpret discovery queries with heterogeneous ontologies and maintain their own ontologies to describe their own services (since semantic interpretation is needed, an ontology matching process should be put in place to provide a common understanding); (ii) it is mandatory to have centralized service registries that provide details on the longevity of ontology references, URIs, etc., and self-contained ontologies within the nodes should be used.

An important contribution of Nedos et al. [48] is the implementation of a *gossip protocol*, which is at the core of discovering and matching heterogeneous ontologies. Each node stores concepts in a buffer. Then, a lightweight ontology-matching mechanism matches those concepts with the ones received. In summary, the discovery process first identifies candidate nodes with compatible ontologies and then uses those nodes to perform the service-matching step.

7.3.3.3 Semantic Web Services in Peer-to-Peer and Grid Environments

Le-Hung et al. [49] present a distributed approach to the semantic discovery of Web services in peer-to-peer-based registries, considering QoS characteristics. They argue that their scheme is scalable, efficient, and reliable. The scalability is achieved through the use of peer-to-peer (P2P) overlays as a service repository network. The considered QoS is determined by the users' feedback for a given service. They emphasize evaluation of the credibility of the reporting users. According to the authors, their QoS model is unique and robust against malicious behavior due to the use of known solutions for trust and reputation management in P2P systems [50,51].

Verma et al. [52] propose the METEOR-S WSDI architecture, an environment for publication of Web services and their discovery in multiple registries. It follows an ontology-based approach that facilitates organization of a registry into domains. Each registry is related to a specific domain using semantics for domain association. It is kept in a custom-made ontology known as the *registries ontology*, which contains the relationships among domains and registries.

Systems configured in a grid environment can deploy data, resources, and so on in a virtual working environment, such as the Internet. The infrastructure exploits heterogeneous resources that can be geographically dispersed. Grid technology lends itself well to implementation of SWSs. Ren et al. [53] present a model for grid-based semantic service discovery with QoS constraints. It uses QoS ranking for matching user-specified preferences rather than the traditional semantic matchmaking capability, as used in engine-based service discovery. The paper presents an efficient QoS model using OWL QoS ontologies to meet the needs of nonfunctional requirements and QoS information collection. It also presents a classification of QoS parameters, grouping them into four categories: (i) *network* QoS parameters (bandwidth, delay, etc.), (ii) *system* QoS parameters (reliability, capacity, etc.), (iii) *task* QoS parameters (memory, CPU usage, response time, etc.), and (iv) *extension* QoS parameters (reputation, security, etc.).

Table 7.1 presents the summary of related work on the Semantic Web, showing the features of the Semantic Web and Web services in different networking environments.

7.3.4 Proposed Semantic QoS Specification Solution for Oppnets

This subsection describes how the concepts of QoS requirements, expressed through the means of the Semantic Web in the service delivery process, can fit into the Oppnet context. We propose a new model for adapting QoS requirements to Oppnets. This model is called *Oppnet SCOW-Q*, which stands for Semantic Capability discOvery With QoS in Oppnets.

Table 7.1 Semantic Web and Web Services in Different Networking Environments

Paradigm/ Proposal	Service Description	Service Specification	Service Discovery
Peer-to-Peer (P2P) Computing			
	• WSMO		
P2P–based registries with QoS Support Le-Hung et al. [49]	• QoS requirements are described as normalized values in a set of triples $\{q_i, n_i, v_i\}$, where: • q_i = a QoS parameter • n_i = the order of importance for q_i • vi = the user's minimal required value	—	• SD, selection and ranking based on the matching of service advertisements considering QoS, trust, and reputation.
METEOR-S WSDI Verma et al. [52]	• WSDL + semantic publication of Web services in UDDI. Other developed algorithms are: SAWS to map concepts in WSDL description to an ontological concept returning the degree of similarity.	• Customized ontology called *Registries Ontology* where ontology/ registry mapping occurs and properties such as QoS are stored.	• In the Operator Service layer of the proposed architecture conventional UDDI query and matching are supported. • It uses SOAP for non-UDDI registry implementation.
Grid Computing			
• Web service Discovery model with QoS constraints	• Service description is transformed onto the OWL-S profile specification	• Parsing of the request specifications to identify concepts and properties.	• Concepts are captured in a vector list (input/ output) for matching purposes. • Extension of a semantic matching algorithm [55,56]

(Continued)

Table 7.1   Semantic Web and Web Services in Different Networking Environments (Continued)

Paradigm/ Proposal	Service Description	Service Specification	Service Discovery
Mobile Ad Hoc Networks			
• Nedos et al. [48]	• RDFS	• OWL-S	• Concepts are represented through a *network representation* • Gossip Protocol • Discovery Query • Syntactical Matching
• Kopena et al. [54]	• OWL-S	• OWL-S	• Random walk service discovery agents, also described as a set of service-monitoring agents • Profile matching
Pervasive Computing			
EASY Ben Mokhtar et al. (2007) [58]	Proprietary EASY-L based on OWL. Support for functional and nonfunctional services	—	—

7.3.4.1 The Oppnet SCOW-Q Model

The proposed model is based on the semantic service discovery process using ontologies. This process is applied to seed Oppnet nodes to perform the *capability discovery* task in Oppnets. The capability discovery approach presented in this model could be implemented in open environments such as the Web, which already has an underlying layer that binds up services using protocols such as HTTP. Moreover, this model is adequate for environments using ad hoc connectivity, such as MANETs or P2P (peer-to-peer) systems.

Common concepts utilized in this model are helper registry and helper advertisements, both based on the concepts known from the semantic service discovery process. *Helper registry* is a list of all identified services that have some degree of functional and nonfunctional (including QoS) characteristics that can be matched with request requirements. *Helper advertisements* describe the capabilities (including services) provided by entities. An *entity* is identified as an outsider node, a seed node, or a helper node. Entities can vary in nature, from a group of PCs on a network to a PDA, or from a satellite connection to an intelligent appliance or a sensor. These participating devices must have a mechanism to advertise their capabilities.

The first entity considered in this model is the Control Center (CC) node. This node has a great deal of interaction with other seed nodes. CC is responsible for having agents that constantly monitor the environment, searching for potential capabilities, maintaining and updating changes in ontologies, and keeping a repository in which a helper registry is maintained. When an Oppnet needs new capabilities, candidate helper nodes are located, and some of them are invited to join the Oppnet as its helpers, providing their capabilities.

A classification of helper nodes is provided later in this section. Helper nodes can have assistants with limited capabilities of their own but also able to locate other nodes and identify their advertised capabilities. These assistants are called *lites* (short for *lightweight helpers*).

Figure 7.3 depicts the Oppnet SCOW-Q model. The establishment and basic operations of Control Center, as well as its collaboration with seed nodes, are shown in the innermost oval. The remaining two ovals, surrounding the innermost oval, connect different types of helpers that can exist during the lifetime of an Oppnet, with the outermost oval connecting only lites.

7.3.4.2 Oppnet SCOW-Q for Seed Nodes and Control Center

Seed nodes are a group of nodes simultaneously deployed to form an initial Oppnet. A distributed Control Center is a subset of the seed nodes that has more management power than other Oppnet nodes. This subsection describes the proposed interaction between seed nodes and the Control Center in the Oppnet SCOW-Q model.

To simplify the Oppnet architecture, we assume that only agents within CC nodes can discover needed helper nodes and identify their capabilities through helper (or service) advertisements. Given that the CC nodes are able to accept or reject candidate helpers, the agent management and the helper registry based on ontologies can reside in them. In an expanded Oppnet, CC continues to carry on its tasks of maintaining up-to-date ontologies and updating the registry of advertised capabilities whenever new advertisements are received (or discovered).

When a new helper joins the Oppnet, a copy of the most current ontology and helper registry is downloaded to the helper from the CC. Similarly, a matchmaking

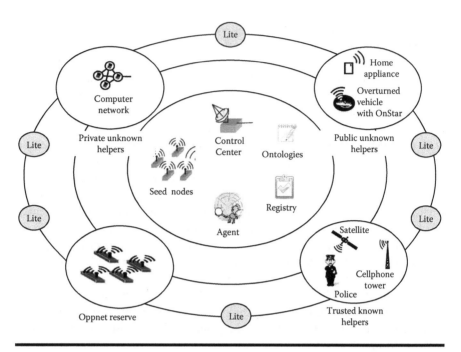

Figure 7.3   The Oppnet SCOW-Q model for semantic capability discovery with QoS parameters.

agent is deployed on the helper, facilitating execution of its own autonomous helper-matching process.

Figure 7.4 shows a process flow for a semantic capability discovery in Oppnet SCOW-Q. The model presented by Nedos et al. [48]—originally for MANETs—is followed in Oppnets. The reason is that the dynamic characteristics of Oppnets are satisfied by the distributed ontology matching, using a gossip protocol ontology dissemination and a walk mechanism to find capability providers. Also, the EASY language and EASY Matching solutions for pervasive computing [58] can be used to more efficiently match the advertised capabilities and the capability requests issued by Oppnet nodes. The efficient mechanism for service discovery provided by EASY is independent of underlying layers and allows for QoS specifications.

Oppnet nodes monitor each other and the candidate helper nodes, keeping track of their behavior. Any anomaly is noticed and recorded. Feedback is submitted to CC for integration at the helper registry level, and for building a reputation database for Oppnet nodes and candidate helpers. Feedback and reputation data can be disseminated via the gossip and walk mechanisms [48].

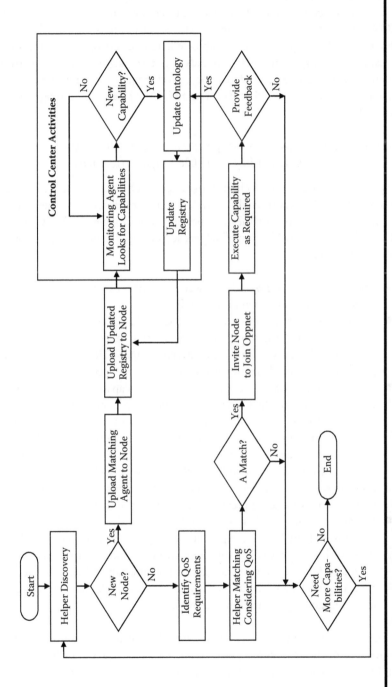

Figure 7.4 Process flows in the Oppnet SCOW-Q Model.

198 ■ *Leszek Lilien et al.*

Another component from EASY applied in Oppnets, and in particular in the CC, is the technique to efficiently organize the helper registry in the Oppnet SCOW-Q model (corresponding to the service repository in EASY). This approach to the helper registry enables efficient insertion of service advertisements, minimizing the impact to already registered services. It also helps reduce the number of matches considered for a given request. In contrast to Oppnets, the semantic reasoning used in EASY to build the classified ontology hierarchies occurs offline.

7.3.4.3 Classification of Helpers According to Their Access, Capability, and QoS Specifications

The pervasive and ubiquitous characteristics of an Oppnet expose it to a large variety of candidate helpers (including lites), which vary in terms of their nature and the extent of the capabilities they can provide. The Oppnet SCOW-Q model provides the following classification of helpers according to their access, capability, and QoS specifications:

- *Public unknown helpers*—Their access policies, capability descriptions, and QoS parameters are open to the public.
- *Private unknown helpers*—Their access policies are private. They do not exchange this information unless a negotiation takes place and an agreement is reached. Helpers of this type advertise capabilities and QoS parameters publicly; however, their use must be preceded by negotiations.
- *Trusted known helpers*—Their access policies, capabilities, and QoS parameters are known to the Oppnet. Seed nodes, or helpers identified as trusted, are, for example, those located within a known trusted environment whose capabilities have already been successfully used by the Oppnet. This category of helpers is based on the concept that establishing a Web of trust (based on social network theories applied to the Semantic Web using ontologies) will result in self-sufficient Oppnet nodes able to manage their network of trusted Oppnet nodes based on previous experience with them [57].
- *Oppnet reservists*—If there is a match between the application category of a reservist and the Oppnet (e.g., both are tagged for emergency applications), they are highly trusted by the Oppnet, even more than the previous group of trusted known helpers. Their access policies, capabilities, and QoS parameters are known a priori to the Oppnet (from the Oppnet reservist directory for the matching application category).

Default QoS parameters can be established to ensure the minimum QoS level for interaction with helpers of a given class in case the requestor does not provide its own QoS specifications. The defaults are saved within service profiles.

7.4 Conclusions

In this chapter we presented Oppnets, a paradigm and a technology proposed for realization of Opportunistic Capability Utilization Networks (also called Class 2 Opportunistic Networks). We delineated components and basic operations of Oppnets and presented use of the service location and planning (SLP) technique for resource utilization with QoS constraints in Oppnets. We illustrated, with the Oppnet SCOW-Q model, the use of Semantic Web technology and its ontologies for specifying QoS requirements in Oppnets.

The SLP mechanism is a novel resource utilization technique, since it not only enables resource utilization but also meets consumer-defined performance QoS requirements of throughput and delay. The ability to capture adequately precise bandwidth and latency characteristics in a resource utilization model is imperative for realistic modeling of scenarios. The SLP model contributes to resource utilization techniques by not only enabling consumers to indicate throughput and delay requirements, but also by taking into account the underlying network parameters determining delays and bandwidths. It has been shown [29,30] that the SLP mechanism can also be used in diverse areas such as mobile computing and Cisco Application-Oriented Networking (AON).

Due to the dynamic, delay-tolerant, and intermittent-connectivity nature of Oppnets, accurate and efficient operation of the SLP model can be computationally intensive. Our future work will entail the use of the SLP model in dynamic Oppnet scenarios, analyzing its performance and scrutinizing the degree of fulfilling of the QoS requirements.

We presented a realization of semantic *capability* discovery in Oppnets (corresponding to semantic *service* discovery). It is an essential activity, necessary to ensure appropriate satisfaction of capability requests in an already deployed Oppnet. We have indicated that the semantic capability discovery process should be enhanced by applying non-functional service specifications in the form of QoS requirements. They assist an Oppnet component (a node or a subnetwork) in the decision-making process of inviting candidate helpers to join an Oppnet. The decision should be based not only on the functional description of the services (or capabilities in general) provided by candidate helpers, but also on the evaluation by other Oppnet components of the known previous behavior of the candidates considered for invitation by the Oppnet.

We are faced with some limitations in this work, such as: (i) the correct use of ontologies, (ii) considerations of the capacity of potential helpers to advertise their capabilities and to retain certain ontological data, and (iii) dependency on signal availability in case of interaction with ad hoc networks. In the future, we will use our results to address the QoS challenges in Oppnets (cf. [70]). Also, applications of the proposed solution in real-life scenarios will lead to an expansion of this work.

References

[1] B. Bhargava , L. Lilien, A. Rosenthal, and M. Winslett. Pervasive trust. *IEEE Intelligent Systems*, 19(5):74–77, September/October 2004.

[2] L. Lilien, Z. H. Kamal, V. Bhuse, and A. Gupta. Opportunistic networks: The concept and research challenges in privacy and security. *Proceedings of the International Workshop on Research Challenges in Security and Privacy for Mobile and Wireless Networks (WSPWN 2006)*, pp. 134–147, Miami, Florida, March 2006.

[3] L. Lilien, Z. H. Kamal, V. Bhuse, and A. Gupta. The concept of opportunistic networks and their research challenges in privacy and security. In *Mobile and wireless network security and privacy*, ed. by K. Makki et al., Springer Science+Business Media, pp. 85–117, Norwell, Massachusetts, 2007.

[4] L. Lilien, Z. H. Kamal, and A. Gupta. Opportunistic networks: Research challenges in specializing the P2P paradigm. In *Proceedings of the 3rd International Workshop on P2P Data Management, Security and Trust (PDMST'06)*, pp. 722–726, Kraków, Poland, September 2006.

[5] L. Lilien. A taxonomy of specialized ad hoc networks and systems for emergency applications. In *Proceedings of the 1st International Workshop on Mobile and Ubiquitous Context Aware Systems and Applications (MUBICA 2007)*, Philadelphia, Pennsylvania, August 2007.

[6] L. Lilien. Developing specialized ad hoc networks: The case of opportunistic networks. Unpublished presentation. *Workshop on Distributed Systems and Networks* at *WWIC 2006 Conference*, Bern, Switzerland, May 2006.

[7] L. Pelusi, A. Passarella, and M. Conti. Opportunistic networking: Data forwarding in disconnected mobile ad hoc networks. *IEEE Communications*, 44(11):134–141, November 2006.

[8] Y. Wang, S. Jain, M. Martonosi, and K. Fall. Erasure-coding based routing for opportunistic networks. In *Proceedings of the ACM Conference of the Special Interest Group on Data Communication (SIGCOMM'05)*, pp. 229–236, Philadelphia, Pennsylvania August 2005.

[9] P. Sistla, O. Wolfson, and B. Xu. Opportunistic data dissemination in mobile peer-to-peer networks. In *Proceedings of the 9th International Symposium on Advances in Spatial and Temporal Databases (SSTD'05)*, pp. 346–363, Angra dos Reis, Brazil, August 2005.

[10] L. Lilien, A. Gupta, and Z. Yang. Opportunistic networks for emergency applications and their standard implementation framework. In *Proceedings of the 1st International Workshop on Next Generation Networks for First Responders and Critical Infrastructure (NetCri07)*, New Orleans, Louisiana, April 2007.

[11] Z. H. Kamal, L. Lilien, A. Gupta, Z. Yang, and M. Batsa. New UMA Paradigm: Class 2 Opportunistic Networks Class 2 Opportunistic Networks. Chapter 17 in *Unlimited mobile access technology: Protocols, architectures, security, standards and applications*, ed. by Y. Zhang, et al., Auerbach Publications, Boca Raton, Florida, pp. 349–392, 2008.

[12] Z. Kamal, A. Gupta, L. Lilien, and Z. Yang. The MicroOppnet tool for collaborative computing experiments with Class 2 Opportunistic Networks. In *Proceedings of the 3rd International Conference on Collaborative Computing: Networking, Applications and Worksharing (CollaborateCom 2007)*, pp. 150–159, White Plains, New York, November 2007.

[13] D. B. Shmoys, C. Swamy, and R. Levi. Facility location with service installation costs. In *Proceedings of the 15th Annual ACM-SIAM Symposium on Discrete Algorithms*, pp. 1088–1097, New Orleans, Louisiana, January 2004.

[14] M. G. C. Resende and R. F. Werneck. *A hybrid multistart heuristic for the uncapacitated facility location problem*. AT&T Labs Research Technical Report, TD-5RELRR, Florham Park, New Jersey, September 2003.

[15] B. Aoun, R. Boutaba, Y. Iraqi, and G. Keyward. Gateway placement optimization in wireless mesh networks with QoS constraints. *IEEE Journal on Selected Areas in Communications*, 24(11):2127–2136, November 2006.

[16] Y. Bejerano. Efficient integration of multihop wireless and wired networks with QoS constraints. *IEEE/ACM Transactions on Networking*, 12(6):1064–1078, December 2004.

[17] P. A. Bonatti, and P. Festa. On optimal service selection. In *Proceedings of the ACM International Conference World Wide Web (WWW'05)*, pp. 530–538, Chiba, Japan, May 10–14, 2005.

[18] J. Cardinal and M. Hoefer. Selfish service installation in networks. In *Proceedings of the International Conference on Internet and Network Economics (WINE'06)*, Vol. 4286, LNCS, pp. 174–185, Springer-Verlag, 2006.

[19] R. Rizzi and M. Rospocher. Covering partially directed graphs with directed paths. *Discrete Mathematics*, 306(13):1390–1404, July 2006.

[20] F. Zhu, M. Mutka, and L. M. Ni. Service discovery in pervasive computing. *IEEE Pervasive Computing*, 4(4):81–90, October–December 2005.

[21] F. A. Chudak and D. P. Williamson. Improved approximation algorithms for capacitated facility location problems. In *Proceedings of the 7th International Conference on Integer Programming and Combinatorial Optimization (IPCO'99)*, pp. 99–113, Graz, Austria, June 1999.

[22] S. Jin and L. Wang. Content and service replication strategies in multi-hop wireless mesh networks. In *Proceedings of the International Symposium on Modeling, Analysis and Simulation of Wireless and Mobile Systems (MSWiM'05)*, pp. 79–86, Montreal, Canada, 2005.

[23] X. Gao, R. Jain, Z. Ramzan, and U. Kozat. Resource optimization for web service composition. In *Proceedings IEEE International Services Computing Conference (SCC)*, Orlando, Florida, 2005. http://ravijainweb.com/mware/gao-webservice-PS-2005-0023-scc05.pdf, accessed August 27, 2008.

[24] H. Janzadeh, K. Fayazbakhsh, B. Bakhshi, and M. Dehghan. A novel incentive-based and hardware-independent cooperation mechanism for MANETs. In *Proceedings of the IEEE Wireless Communications and Networking Conference (WCNC 2008)*, pp. 2462–2467, Las Vegas, Nevada, March–April 2008.

[25] Cisco Systems Inc., Introducing Cisco Application-Oriented Networking—A CIO Brief, http://www.cisco.com, 2006, accessed October 5, 2007.

[26] H. M. Kim, A. Sengupta, and J. Evermann. MOQ: Web services ontologies for QoS and general quality evaluations, pp. 195–200, http://is2.lse.ac.uk/asp/aspecis/20050134.pdf, accessed August 27, 2008.

[27] T. Gagnes, T. Plagemann, and E. Munthe-Kaas. A conceptual service discovery architecture for semantic web services in dynamic environments. In *Proceedings of the Workshops at the 22nd International Conference on Data Engineering*, p. 74, 2006.

[28] O. V. Drugan, T. Plagemann, and E. Munthe-Kaas. Building resource aware middleware services over MANET for rescue and emergency applications. In *Proceedings of the 16th Annual IEEE International Symposium on Personal Indoor and Mobile Radio Communications*, pp. 816–820. Berlin, Germany, September 2005.

[29] Z. H. Kamal, A. Al-Fuqaha, and A. Gupta. A service location problem with QoS constraints. In *Proceedings of the 2007 International Conference on Wireless Communications and Mobile Computing (IWCMC'07)*, pp. 641–646, Honolulu, Hawaii, August 2007.

[30] Z. H. Kamal, A. Al-Fuqaha, and A. Gupta. Using Lagrangean relaxation for service location planning with QoS constraints in large-scale networks. In *Proceedings of the IEEE International Conference on Communications (ICC 2008)*, pp. 424–428, Beijing, China, May 2008.

[31] T. Berners-Lee, J. A. Hendler, and O. Lassila. The Semantic Web. *Scientific American*, 284(5):34–43, May 2001.

[32] F. Baader, D. Calvanese, D. L. McGuinness, D. Nardi, and P.F. Patel-Schneider. *The description logic handbook: Theory, implementation, and applications*. Cambridge, U.K., Cambridge University Press, 2003.

[33] The W3C Organization. Web services architecture, 2004, http://www.w3.org/TR/ws-arch/, accessed July 27, 2008.

[34] C. Kiss. Composite capability/preference profiles (CC/PP): Structure and vocabularies 2.0, W3C Working Draft, April 30, 2007, http://www.w3.org/TR/2007/WD-CCPP-struct-vocab2-20070430, accessed August 15, 2008.

[35] The OWL Service Coalition. OWL-S semantic mark-up for web services, 2003, http://www.daml.org/services/owl-s/1.0/owl-s.html, accessed August 1, 2008.

[36] The W3C Organization. Web Services Description Language (WSDL) 1.1, 2003, http://www.w3.org/TR/2001/NOTE-wsdl-20010315, accessed July 27, 2008.

[37] The W3C Organization. Web Services Description Language (WSDL) Version 2.0 Part 1: Core language, 2007, http://www.w3.org/TR/wsdl20/, accessed July 27, 2008.

[38] Terms and definitions related to quality service and network performance including dependability, ITU-T Recommendation E.800, August 1994.

[39] ISO/IEC JTC 1: Information technology—Open Distributed Processing—Reference Model: Overview, ISO-Standard ISO/IEC 10746-1, 1998.

[40] M. Anbazhagan and A. Nagarajan. Understanding quality of service for Web services, developerWorks,® SOA and Web services/Technical library, IBM, 2002, http://www-106.ibm.com/developerworks/library/ws-quality.html, accessed July 15, 2008.

[41] D. Tsesmetzis, I. G. Roussaki, I. V. Papaioannou, and M. E. Anagnostou. QoS awareness support in web-service semantics. In *Proceedings of the Advanced International Conference on Telecommunications & International Conference on Internet and Web Applications and Services (AICT/ICIW)*, February 2006, p. 128 (7 pages), http://citeseerx.ist.psu.edu/viewdoc/similar;jsessionid=50D45B6DA15C0BD017EE3F6E3A7D E77B?doi=10.1.1.86.8678&type=ab, accessed September 1, 2008.

[42] M. Zghaibeh and F. C. Harmantzis. Lottery-based pricing scheme for peer-to-peer networks. In *Proceedings of the IEEE International Conference on Communications (ICC)*, 2:903–908, June 2006.

[43] C. Boldrini, M. Conti, and A. Passarella. Context resource awareness in opportunistic network data dissemination. In *Proceedings of the International Symposium. World of Wireless, Mobile and Multimedia Networks (WoWMoM'08)*, pp. 1–6, Newport Beach, California, June 2008.

[44] S. Ran. A model for web services discovery with QoS. *SIGecom Exchanges*, 4(1):1–10, 2003.

[45] J. Gaber. Spontaneous emergence model for pervasive environments. In *Proceedings of the IEEE Globecom Workshops*, pp. 1–4, November 2007.

[46] D. Chakraborty, A. Joshi, Y. Yesha, and T. Finin. Toward distributed service discovery in pervasive computing environments. *IEEE Transactions on Mobile Computing*, 5(2):97–112, 2006.

[47] S. Ben Mokhtar, N. Georgantas, and V. Issarny. COCOA: Conversation-based service composition in pervasive computing environments with QoS support. *Journal of Systems and Software*, 80(12-Special Issue):1941–1955, 2007.

[48] A. Nedos, K. Singh, and S. Clarke. Mobile ad hoc services: Semantic service discovery in mobile ad hoc networks. In *Proceedings of the International Conference on Service Oriented Computing (ICSOC 2006)*, *LNCS* 4294, pp. 90–103. 2006.

[49] V. Le-Hung, M. Hauswirth, and K. Aberer. Towards P2P-based semantic web service discovery with QoS support. *Lecture Notes in Computer Science* 3812:18–31, 2006.

[50] Z. Despotovic and K. Aberer. Possibilities for managing trust in P2P networks, Technical Report IC200484, Swiss Federal Institute of Technology at Lausanne (EPFL), Switzerland, November 2004.

[51] A. Jøsang, R. Ismail, and C. Boyd. A survey of trust and reputation systems for online service provision. *Decision Support Systems*, 43(2):618–644, March 2007.

[52] K. Verma, K. Sivashanmugam, A. Sheth, A. Patil, S. Oundhakar, and J. Miller. METEOR-S WSDI: A scalable P2P infrastructure of registries for semantic publication and discovery of web services. *Information Technology and Management*, 6(1):17–39, 2005.

[53] K. Ren, J. Chen, T. Chen, J. Song, and N. Xiao. Grid-based Semantic Web service discovery model with QoS constraints. In *Proceedings of the IEEE 3rd International Conference on Semantics, Knowledge and Grid*, pp. 479–482, October 2007.

[54] J. Kopena, E. Sultanik, G. Naik, I. Howley, M. Peysakhov, V. A. Cicirello, M. Kam, and W. Regli. Service-based computing on MANETs: Enabling dynamic interoperability of first responders. *IEEE Intelligent Systems*, 20(5): 17–25, September–October 2005.

[55] M. Paolucci, K. Sycara, and T. Kawamura. Delivering Semantic Web services. In *Proceedings of the 12th International World Wide Web Conference (WWW2003)*, pp. 34–41, Budapest, Hungary, 2003.

[56] N. Srinivasan, M. Paolucci, and K. Sycara. An efficient algorithm for OWL-S based semantic search in UDDI. In *Semantic web service and web process composition, Lecture Notes in Computer Science* 3387:96–100, 2005.

[57] J. Golbeck, B. Parsia, and J. Hendler. Trust networks on the Semantic Web. *Lecture Notes in Artificial Intelligence (Subseries of Lecture Notes in Computer Science)*, 2782:238–249, 2003.

[58] S. Ben Mokhtar, D. Preuveneers, N, Georgantas, V. Issarny, and Y. Berbers. EASY: Efficient Semantic Service Discovery in pervasive computing environments with QoS and context support. *Journal of Systems and Software*, 81(5):785–808, 2007.

[59] The DARPA Agent Markup Language (DAML) Program, Semantic Web Services, 2004, http://www.daml.org/services/, accessed August 12, 2008.

[60] The ESSI WSMO working group, http://www.wsmo.org/, accessed August 12, 2008.

[61] LG R&D News, LG Electronics, http://www.lge.com/about/rnd_news/detail/2275_7.jhtml, accessed September 2, 2008.

[62] Pervasive Computing, SearchNetworking.com, Definitions, http://searchnetworking.techtarget.com/sDefinition/0,,sid7_gci759337,00.html, accessed September 7, 2008.

[63] P. Juang, H. Oki, Y. Wang, M. Martonosi, L.-S. Peh, and D. Rubenstein. Energy-efficient computing for wildlife tracking: Design tradeoffs and early experiences with ZebraNet. In *Proceedings of the ACM Conference on Architectural Support for Programming Languages and Operating Systems (ASPLOS)*, pp. 96–107, San Jose, California, October 2002.

[64] H. Yoon, J. Kim, F. Tan, and R. Hsieh. On-demand video streaming in mobile opportunistic networks. In *Proceedings of the IEEE 6th International Conference on Pervasive Computing and Communications (PerCom)*, pp. 80–89, Hong Kong, March 2008.

[65] Haggle: A European Union funded project in situated and autonomic communications, Haggle Project, http://www.haggleproject.org/index.php/Main_Page, accessed September 8, 2008.

[66] The iClouds Project: Opportunistic communication among people, http://iclouds.tk.informatik.tu-darmstadt.de/, accessed September 8, 2008.

[67] C. Peng, H. Zheng, and B. Y. Zhao. Utilization and fairness in spectrum assignment for opportunistic spectrum access. *Mobile Networks and Applications*, 11(4):555–576, August 2006.

[68] N. B. Chang and M. Liu. Competitive analysis of opportunistic spectrum access strategies. In *Proceedings of the IEEE 27th Conference on Computer Communications* at *INFOCOM 08*, pp. 1535–1542, Phoenix, Arizona, April 2008.

[69] Y. Liu, A. H. H. Ngu, and L. Zeng. QoS computation and policing in dynamic web service selection. In *Proceedings of the 13th International World Wide Web Conference (WWW 2004)*, pp. 798–805, New York, May 2004.

[70] S. Marwaha, J. Indulska, and M. Portmann. Challenges and recent advances in QoS provisioning in wireless mesh networks: A survey. In *Proceedings of the IEEE 8th International Conference on Computer and Information Technology*, pp. 618–623, Sydney, Australia, July 2008.

Chapter 8

Effective File Transfer in Mobile Opportunistic Networks

Ling-Jyh Chen
Institute of Information Science, Academia Sinica

Ting-Kai Huang
University of California at Riverside

Contents

The last few years have seen an impressive growth in computer network applications. One striking success in this area has been the Internet, which has successfully accelerated the dissemination of information and knowledge by overcoming geographic boundaries. Wireless technologies represent another orthogonal area of growth, in both wide-area applications like 2.5G/3G and local area applications like 802.11b/g and Bluetooth. As wireless technologies continue to extend into every part of our working and living environments, it is becoming increasingly desirable to have a solution that can provide effective data transfer on the go.

Proper handling of mobility is the key to the success of mobile networking applications, and a great amount of research effort has been invested to facilitate mobile applications in the last few years [9,23,43,44,49,51,54,56]. Generally, these approaches can be categorized into three types: network layer–based approaches [23,49], transport layer–based approaches [9,44,56], and proxy-based approaches [43,51,54]. The common goal of these approaches is to maintain the Internet connectivity for mobile users even when they perform vertical handoffs between different networks. However, the major shortcoming of these approaches is that they are all Internet based and thus limited in providing FTP service when the Internet connection is not always available.

Fortunately, as recent studies have reported that intermittent network connection is inevitable for mobile users on a daily basis [14,16], Grossglauser and Tse [28] have also shown that network capacity can be increased dramatically by exploiting node mobility as a type of multiuser diversity. In other words, opportunistic ad hoc connections can be useful for extending the coverage of wireless communications.

Several approaches have been proposed to allow mobile users to request and transfer data on the go [6,30,34,45,46,47,48,55]. The approaches are either cache based or Infostation based. Cache-based approaches facilitate mobile file transfer by prefetching popularly requested files (e.g., commercial ads, movie trailers, and song previews) to a local storage. Infostation-based approaches, on the other hand, support on-demand FTP requests by deploying dedicated servers as bridges between the Internet and mobile networks. However, the capability of these approaches for mobile file transfer is limited because they are basically centralized and fail to exploit the diversity of network mobility.

In this chapter, we present a peer-to-peer approach, called M-FTP, for mobile file transfer applications. In addition to combining the strengths of cache-based approaches (i.e., prefetching the most likely requested files to a local storage) and Infostation-based approaches (i.e., allowing on-demand FTP requests), the M-FTP scheme implements a *collaborative forwarding* algorithm to further utilize opportunistic ad hoc connections and spare storage in the network. Using simulations as well as real-world mobility traces, we evaluate the M-FTP scheme in terms of service ratio and traffic overhead. The results show that the scheme significantly outperforms previous approaches in all test cases, while its traffic overhead remains moderate.

The remainder of this chapter is organized as follows. In Section 8.1 we review related work on mobile file transfer and opportunistic network routing in mobile networks. In Section 8.2 we describe the M-FTP scheme and the collaborative for-warding feature used to disseminate data in mobile networks. Section 8.3 presents a comprehensive set of simulation results, which we analyze and explain in detail. We then present our conclusions in Section 8.4.

8.1 Related Work

8.1.1 Mobile Data Transfer

Data transfer in mobile networks has been researched for a number of years, and sev-eral approaches have been proposed [6,30,34,45,46,47,48,55]. The approaches can be roughly classified into two types: cache-based and Infostation-based approaches.

Basically, cache-based approaches automatically download those files that are considered likely to be requested in the near future. This is done in either a *push*-based [6,46,55] or a *pull*-based [34,48] fashion. More precisely, when push-based approaches are used, the content provider automatically supplies a mobile user with popularly requested files as long as he/she is connected to the Internet and has free storage space. In contrast, under pull-based approaches, the mobile device auto-matically pulls (prefetches) files (using its own content selection algorithm) without the FTP requests being issued manually by mobile users. There is a trade-off in these approaches; they incur a tremendous storage overhead in return for the per-formance gain (i.e., the more files they cache in the local storage, the greater the likelihood that they will be able to serve the next FTP request without consum-ing extra Internet bandwidth). However, this is considered infeasible for emerging power/storage-constrained handheld devices. In addition, these approaches only allow Internet-capable users to download files, but they do not provide a way for users without Internet capabilities to download.

Infostation-based approaches provide Internet access for mobile users by installing Infostations such as Mobile Ad hoc Networks (MANET) or delay- or disruption-tolerant networks (DTN) to act as bridges between the Internet and mobile networks [27,52]. The advantage of such approaches is that they allow mobile users to obtain/publish data on the Internet via Infostations by using local wireless connections (e.g., WiFi and Bluetooth). Hence, mobile file transfer is possible, even when mobile users are not directly connected to the Internet. For example, [30] proposes a solution called *Mobile Hotspots*, which provide mobile Internet access in railway systems; [45] proposes the *SPAWN* scheme, which provides coopera-tive content delivery via Internet gateways to the inner vehicular ad hoc networks (VANET); and [47] proposes a Bundle Routers (BR)-based approach that deploys BRs as Infostations to separate the Internet and DTNs. Unlike previous schemes, the SPAWN scheme and the BR-based scheme require mobile users' collaboration to forward data in a multihop and store-carry-and-forward fashion. It also allows users to access Internet data, even if they cannot locate an Infostation. However, the major shortcoming of the above approaches is that they require dedicated

Infostations (i.e., gateways or BRs), which act as gateways to the Internet; thus, they suffer from scalability and single-point-of-failure problems.

8.1.2 Opportunistic Network Routing

Replication is the most popular design choice for opportunistic routing schemes. For instance, the *epidemic routing* scheme [57] sends identical copies of a message simultaneously over multiple paths to mitigate the effects of a single path failure; thus, it increases the possibility of successful message delivery. However, flooding a network with duplicate data tends to be very costly in terms of traffic overhead and energy consumption.

To address the problem of excess traffic overhead caused by flooding, Harras et al. [29] proposed a *controlled flooding* scheme that reduces the flooding overhead while maintaining reliable message delivery. To control flooding, the scheme uses three parameters: *willingness probability, time to live,* and *kill time*. Additionally, after a message has been delivered successfully, a *passive cure* is generated to "heal" network nodes that have been "infected" by the message. Controlled flooding substantially reduces the network overhead by preventing the excess traffic overhead problem, while maintaining reliable data delivery.

Node mobility also impacts on the effectiveness of opportunistic routing schemes. Previous studies have shown that if the network mobility differs from that of well-known random waypoint mobility models (e.g., the Pursue Mobility Model [15] or the Reference Point Group Mobility Model [31]), the overhead carried by epidemic- and/or flooding-based routing schemes can be reduced by considering node mobility. For instance, the *probabilistic routing* scheme [41] calculates the *delivery predictability* from a node to a particular destination node based on the observed contact history and forwards a message to its neighboring node if and only if that node has a higher delivery predictability value. Leguay et al. [38] extended the scheme by taking the *mobility pattern* into account—that is, a message is forwarded to a neighbor node if and only if that node has a mobility pattern similar to that of the destination node. The results reported in [38,39] show that the extended mobility pattern scheme is more effective than previous schemes.

Another class of opportunistic network routing schemes is based on encoding techniques, which transform a message into a different format prior to transmission. For example, to reduce the number of transmissions required in a network, an integration of network coding and epidemic routing techniques was proposed in [60]. Meanwhile, [59] suggested combining erasure coding and the simple replication-based routing method to improve data delivery for cases with the worst delay performance in opportunistic networks.

Following the concept of erasure coding-based data forwarding [59], Y. Liao et al. proposed an estimation-based erasure-coding routing scheme (EBEC) that adapts the delivery of erasure-coded blocks by using the Average Contact Frequency (ACF) estimate [40]. In addition, [18] proposed a hybrid scheme called HEC, which combines the strength of erasure coding and the advantages of *aggressive forwarding.* The HEC scheme has been further enhanced by employing techniques like sequential forwarding (i.e., HEC-SF) [19], probabilistic forwarding (i.e., HEC-PF) [17], full interleaving (i.e., HEC-FI) [19], and block-based interleaving (i.e., HEC-BI) [19].

8.2 The Proposed Approach: M-FTP

In this section, we present a peer-to-peer approach, called M-FTP, for mobile file transfer in opportunistic people networks. The approach combines the strengths of cache-based and Infostation-based approaches with peer-to-peer networking concepts. As a result, it is better able to cope with the intermittent network connectivity caused by mobility and can therefore provide better mobile FTP services. We describe M-FTP in detail in the following subsections.

8.2.1 M-FTP Architecture

There are two types of participating peers in the M-FTP system: *gateway peers* (GPs) and *vanilla peers* (VPs). GPs are connected to the Internet directly (by using, for example, general packet radio service (GPRS), Universal Mobile Telecommunications Systems (UMTS), WiMAX, or WiFi/Bluetooth via Internet access points). VPs are peers that do not have Internet access, but they have local wireless connection capabilities (by using, for example, WiFi, Bluetooth, or infrared via the ad hoc connection mode). Note that a mobile peer may switch from the GP mode to the VP mode (and vice versa) if it temporarily loses (or recovers) its Internet connection (e.g., when entering/leaving an elevator or a tunnel).

In the M-FTP system, there are two cases when a peer A requests to download a file: if A is a GP, he can download the file himself immediately; otherwise (i.e., A is a VP), A forwards his request, with a replication factor f (i.e., f copies of the request are input to the network), to the first f peers he meets in the network. Of course, the larger the value of f, the higher the number of participating peers that will be aware of A's request; however, the traffic and storage overhead also increase linearly as f increases.

The proposed M-FTP system is then applied as follows (see Figure 8.1). Suppose B is another mobile peer that receives A's request. There are two cases:

1. *If B is a GP*, he immediately downloads the requested file from the Internet and forwards the file to A if they are directly connected (i.e., by the *direct forwarding* algorithm). Next, B disseminates the file to the mobile network using the *collaborative forwarding* algorithm, which we will discuss in the next subsection. Note that the objective of the collaborative forwarding algorithm is to cache files previously requested by peers in the network. This allows the M-FTP system to reduce redundant downloading when multiple peers request the same file. Of course, proper buffer management is required to

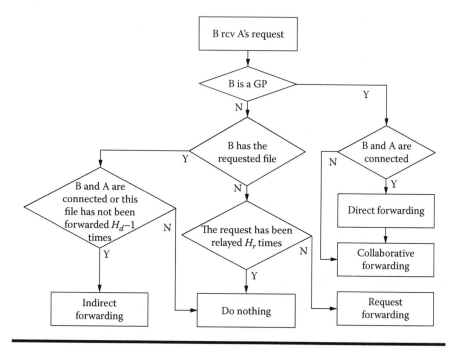

Figure 8.1   Illustration of the request process algorithm of the M-FTP scheme.

further improve the performance of the M-FTP system. We defer a detailed discussion and evaluation of this issue to a future work.

2. *If B is a VP*, he first checks his local storage to determine whether the requested file has been cached and then implements one of the following two options:

   a. *If B has the file requested by A*, he forwards it to *A* if they are directly connected or the file has not been forwarded $H_d - 1$ times (i.e., by the *indirect forwarding* algorithm); otherwise, he does nothing.* Note that the indirect forwarding phase is slightly different from the direct forwarding phase, since *B* may only have a portion of the requested file (which depends on the underlying collaborative forwarding algorithm), whereas the complete file is forwarded in the direct forwarding phase.

   b. *If B does not have the file requested by A*, he forwards *A*'s request to his next encountered peer so long as the request has not been forwarded (from *A* to *B*) more than H_r times (i.e., the *request forwarding* algorithm); otherwise, he does nothing.

Additionally, the M-FTP system prioritizes the data transmissions such that direct forwarding has the highest priority, followed by request forwarding. Then,

---

* In order to reduce the overall traffic overhead in the network, each data block is allowed to be forwarded at most H_d times in the M-FTP scheme.

the other types of transmissions are processed on a first-come-first-served basis. The reason is that direct forwarding can complete an FTP session and thus shorten the service time of that transmitted page. Similarly, request forwarding propagates FTP requests to the network, which increases the probability that the requests will reach GPs and thus be cached in the network.

8.2.2 Collaborative Forwarding Algorithm

In this study, we incorporate the Probabilistic Routing (PRoPHET) algorithm [41] to provide collaborative forwarding for the M-FTP system. The PRoPHET scheme is based on the *epidemic routing scheme* [57], where each node blindly floods blocks of data to its neighbors as long as they are within its transmission range. In contrast, the PRoPHET scheme forwards data blocks to a newly encountered node only if that node has a higher likelihood of successfully forwarding the message (called the *delivery predictability*) to the destination peer than to the current peer.

More specifically, the PRoPHET scheme is designed based on an assumption that *the node that is often encountered has a higher delivery predictability than the others*. Thus, when peer A encounters peer B, the PRoPHET scheme immediately updates the corresponding *delivery predictability* $(P_{A,B})$ using Equation (8.1), where $P_{encounter}$ is a constant and $0 \leq P_{encounter} \leq 1$.

$$P_{A,B}^{new} = P_{A,B}^{old} + \left(1 - P_{A,B}^{old}\right) \times P_{encounter}. \tag{8.1}$$

Moreover, if A and B do not encounter each other during a period of time, they are less likely to be good forwarders to each other, and the PRoPHET scheme gradually reduces the delivery predictability using Equation (8.2), where K is the number of time units since the last time A and B were encountered, and γ is an aging constant $(0 \leq \gamma \leq 1)$.

$$P_{A,B}^{new} = P_{A,B}^{old} \times \gamma^{K}. \tag{8.2}$$

Additionally, the PRoPHET scheme considers the *transitive property* when calculating the delivery predictability. More precisely, the delivery predictability $P_{A,B}$ comprises two parts—the direct message transfer (i.e., A transmits data to B directly) and the indirect message transfer (i.e., A has to rely on other nodes relaying data to B). The delivery predictability calculation is shown in Equation (8.3), where C is the relay node and b is a scaling constant determining the impact of indirect message transfer on the overall delivery predictability.

$$P_{A,B}^{new} = P_{A,B}^{old} + (1 - P_{A,B}^{old}) \times P_{A,C} \times P_{C,B} \times \beta. \tag{8.3}$$

Table 8.1   Music Types of the Selected iTunes Songs

Music Type	Number of Songs
Pop	22
Hip-Hop/Rap	29
Rock	12
Alternative	10
Country	10
Soundtrack	9
R&B/Soul	6
Bandas sonoras	1
iTunes Latino	1

Note that the PRoPHET scheme forwards data blocks to a newly encountered node only if that node has a greater delivery predictability to the destination peer than the current peer. Consequently, in comparison with the ordinary epidemic routing approach that blindly floods the network, the traffic overhead can be greatly reduced.

8.3 Evaluation

We now evaluate the performance, in terms of the service ratio and traffic overhead, of the M-FTP scheme and the Mobile Hotspots scheme using a Java-based simulator called DTNSIM [4]. As mentioned earlier, Mobile Hotspots is an Infostation-based approach with mobile Internet gateways, but without a collaborative forwarding feature (i.e., a mobile user can only download files if he encounters one of the gateways).

In each simulation run, we randomly select γ mobile peers as GPs (with unlimited Internet connection bandwidth) and 20% of the other peers (i.e., VPs) as FTP requesters. For simplicity, we focus on music downloading in this study, and, to be realistic,* the FTP requests are based on the distribution of the top 100 requested iTunes [1] songs as reported in the iTunes store on September 7, 2007. The 100 selected iTunes songs are in Advanced Audio Coding (AAC) format [2], and the average file size is about 3.03 MB. Table 8.1 shows the distribution of the music types of the selected songs.

---

* We assume that the file that has been frequently downloaded in the history is more likely to be requested in the near future [7,20].

We assume that FTP requesters only issue FTP requests in the first 10% of the simulation time, with a Poisson rate of 1,800 seconds/request. We also assume that data transmission between mobile peers is wireless at a fixed rate of 2 Mbps, and that each FTP request can be relayed at most 2 hops ($H_r = 2$). For the collaborative forwarding scheme (i.e., PRoPHET), the $P_{encounter}$ parameter is set to 0.75, the aging constant γ is set to 0.98, the transitive property parameter b is set to 0.25, the buffer size of each mobile peer is unlimited, and each data block is allowed to be forwarded at most 5 times ($H_d = 5$). All the simulation results presented in this section are based on the average performance of 200 simulation runs.

8.3.1 Evaluation Scenarios

We evaluate three network scenarios based on realistic wireless network traces—the iMote [3], UCSD [5], and IBM [10] traces, which are publicly available for research purposes and correspond to the opportunistic people networks of conference, campus, and enterprise scenarios, respectively. Table 8.2 outlines the basic properties of the network scenarios.

The iMote trace is a human mobility trace collected at the 2005 IEEE Infocom conference. It was aggregated from 41 Bluetooth-based iMote devices distributed to the student attendees for the duration of the three-day conference. Each iMote device was preconfigured to periodically broadcast query packets to find other Bluetooth devices within range and record the devices that responded to the queries. In addition to the distributed iMote devices, another 233 devices were recorded in the trace. They may have been other Bluetooth-enabled devices (e.g., PDAs, cell phones, or headsets) used during the conference. For simplicity, we assume there is a network contact between two Bluetooth devices if there are query-and-response interactions between them.

Table 8.2   The Properties of the Three Network Scenarios

Trace Name	iMote	UCSD	IBM
Device	iMote	PDA	Laptop
Network Type	Bluetooth	WiFi	WiFi
Duration (days)	3	77	29
Devices Participating	274	273	1,366
Number of Contacts	28,217	195,364	1,176,264
Average Number of Contacts/Pair/Day	0.25148	0.06834	0.04548

The University of California, San Diego (UCSD) trace is client based and records the availability of WiFi-based access points (APs) for each participating portable device (e.g., PDAs and laptops) on the UCSD campus. The network trace covers a 2.5-month period, and there are 273 participating devices. Similar to [16,18,33], we assume that two participating devices in ad hoc mode encounter a communication opportunity (i.e., a network contact) if they are associated with the same AP at the same time.

The IBM trace is also a client-based trace collected by 177 Wi-Fi access points located in three buildings of the IBM Watson Research Center. The trace is about 29 days long, and 1,366 unique MAC addresses are recorded in the trace. Similar to [10], we assume that each unique MAC address corresponds to a user, even though it is possible for a single user to have more than one MAC address or for users to trade cards with each other. Again, we assume that two users encounter a network contact if they are both associated with the same AP at the same time.

8.3.2 Evaluation I: Service Ratio

Here, we evaluate the service ratio performance of the M-FTP scheme and the Mobile Hotspots scheme. Figures 8.2, 8.3, and 8.4 show the experiment results with various γ values in the iMote and UCSD scenarios in cumulative distribution function (CDF) curves.

From Figures 8.2, 8.3, and 8.4, we observe that the proposed M-FTP scheme outperforms the Mobile Hotspots scheme. More specifically, the M-FTP scheme is able to transmit more than 15% of files in all test cases. This confirms our intuition

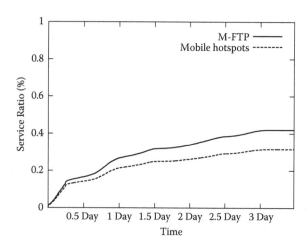

Figure 8.2 Comparison of the service ratio performance of the M-FTP and Mobile Hotspots schemes with various numbers of GPs in the iMote scenario (γ = 20, 40, 60%).

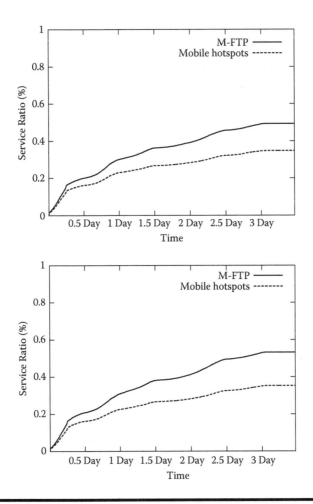

Figure 8.2 (Continued)

that collaborative forwarding can utilize opportunistic connections and thus better exploit the diversity of network mobility. We also observe that the service ratio improves as the value of γ increases, especially in the iMote scenario. The reason is that the iMote scenario's mobility trace was collected by 41 iMote devices, but it did not record network contacts among the other 233 external devices. Consequently, iMote devices generally have a much greater number of network contacts than external devices. Moreover, the Mobile Hotspots scheme only performs effectively if most GPs are iMote devices (otherwise, the GPs cannot contribute much since they seldom encounter VPs in the network scenario). Thus, the higher the value of γ, the greater the likelihood that most GPs in the network will be iMote devices.

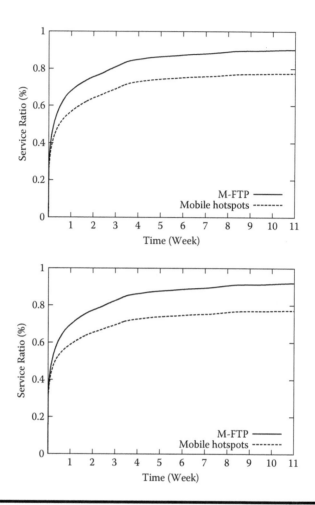

Figure 8.3   Comparison of the service ratio performance of the M-FTP and Mobile Hotspots schemes with various numbers of GPs in the UCSD scenario (γ = 20, 40, 60%).

On the other hand, if each mobile peer has a similar number of network contacts (as in the UCSD scenario), the results show that the service ratio performance is consistent with the changes in the γ values for both schemes. Once again, M-FTP is superior to Mobile Hotspots in all test cases.

8.3.3 Evaluation II: With Free Riders

Free riding is a type of selfish behavior [8,32] that has been extensively studied in the problem of peer-to-peer networking in recent years [11,26,35,50,58]. In the proposed system, free riders may take advantage of the system by issuing numerous

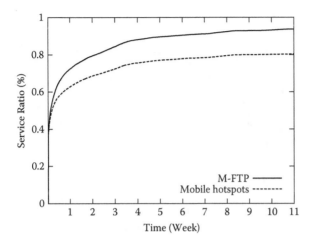

Figure 8.3   (Continued)

requests for Internet content, but refuse to help download/relay messages that are requested by other peers. Consequently, free riders save more of their storage and energy than honest peers, while the system has to pay the cost in terms of delivery ratio performance due to the decreased level of collaboration.

Figure 8.5 demonstrates the delivery ratio performance of the three routing schemes with various percentages of free riders in the network. It is obvious that the delivery ratio performance degrades as the percentage of free riders increases for

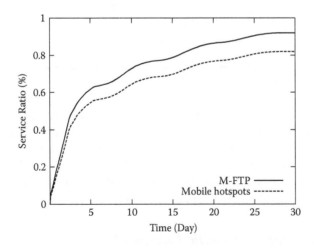

Figure 8.4 Comparison of the service ratio performance of the M-FTP and Mobile Hotspots schemes with various numbers of GPs in the IBM scenario (γ = 20, 40, 60%).

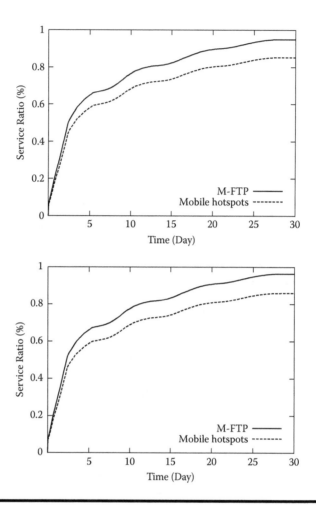

Figure 8.4   (Continued)

all test cases. The results indicate that *free riders are very harmful for data transmission in opportunistic networks*. A potential solution to this issue is to incorporate an incentive-based mechanism into the proposed schemes. Such a mechanism should be able to encourage network peers to contribute their resources and discourage them from engaging in malicious behavior [36]. Consequently, the proposed M-FTP scheme could benefit from the tight collaboration among network peers.

8.3.4 Evaluation III: Traffic Overhead

Next, we evaluate the traffic overhead of M-FTP and Mobile Hotspots. The simulation settings are the same as those in the previous subsection, and the results are based on the average traffic overhead of 200 simulations. Table 8.3 shows the

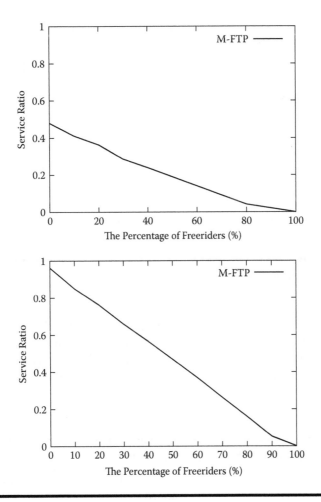

Figure 8.5 **Comparison of the service ratio performance of the M-FTP scheme with various percentages of free riders in the network in the iMote, UCSD, and IBM scenarios ($\gamma = 40\%$).**

results where the normalized overhead is derived by taking the ratio of the traffic overhead of the M-FTP scheme over the Mobile Hotspots scheme.

The results in Table 8.3 show that the normalized traffic overhead is about 3.5 in the iMote scenario and 5.5 in the UCSD scenario. The reason is because the M-FTP scheme disseminates FTP requests and downloaded files to multiple peers to take more aggressive advantage of the diversity of network mobility. Of course, the traffic overhead can be adjusted by tuning the parameters (i.e., H_r and H_d), replacing the collaborative forwarding algorithm, or implementing overhead reduction strategies (e.g., *explicit ACKs* [13], *passive cure* [29], *time to live* [29],

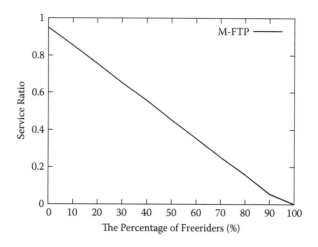

Figure 8.5 **(Continued)**

kill time [29], and *reverse path forwarding* [22]). Generally, the more replicated data stored in the network, the better the service ratio achieved by the M-FTP scheme. The trade-off should be based on the required service ratio and the available buffer sizes. We defer a detailed discussion and evaluation of this issue to a future work.

Table 8.3 Comparison of the Traffic Overhead Performance of the M-FTP and Mobile Hotspots Schemes (Units: Mbytes).

	γ	M-FTP	Mobile Hotspots	Normalized Overhead
	20%	22,170	5,866	3.78
iMote	40%	23,932	6,613	3.62
	60%	24,696	7,197	3.43
	20%	1,425,943	269,834	5.28
UCSD	40%	1,510,094	261,653	5.77
	60%	1,535,310	261,820	5.86
	20%	1,562,502	545,449	2.86
IBM	40%	1,538,177	556,225	2.77
	60%	1,432,240	551,607	2.60

8.3.5 Discussion

We have demonstrated that the M-FTP scheme has the potential to provide effective and efficient mobile data transfer services in opportunistic people networks. The success of the M-FTP scheme depends to a great extent on close collaboration among network participants; however, on the downside, the M-FTP scheme may become very vulnerable if there are uncooperative peers in the network. Here, we discuss the major challenges faced by the M-FTP scheme and present potential solutions to these issues.

First, based on the different levels of malicious behavior, the M-FTP scheme may suffer four types of network attacks—*dropping all packets*, *flooding*, *routing information falsification*, and *ACK counterfeiting* [12]. More precisely, while the *dropping all packets* attack intentionally drops all received packets without forwarding to other peers (a.k.a. black holes [25]), the *flooding* attack continuously sends fake data from any node to any node to drain available network resources (e.g., the energy and buffer of network peers). The *routing information falsification* attack creates erroneous routing information that may cause the network to delay or lose messages altogether, and the *ACK counterfeiting* attack propagates false acknowledgments of in-transit packets that can cut off possibly viable paths to the destination. A potential solution to the issues of uncooperative behavior is to incorporate an incentive-based mechanism (e.g., [24,36]) into the proposed schemes. Such a mechanism should be able to encourage network peers to contribute their resources and discourage them from engaging in malicious behavior [36]. Consequently, the M-FTP scheme would benefit from the tight collaboration among network peers.

Second, since the M-FTP scheme operates on a store-carry-and-forward basis [53], it tends to buffer large amounts of data in the network. Moreover, the required network storage space increases dramatically as the number of file transfer requests increases. Consequently, buffer overflow may occur frequently, such that some in-transit packets have to be dropped, which may substantially degrade the performance of the M-FTP scheme in terms of the delivery ratio. For simplicity, we assume in this chapter that each peer has an infinite buffer size. However, ideally, the solution should consider the nature of the employed routing algorithm and apply an appropriate queuing policy to manage the network storage space [21,37,42].

Finally, a requested data file is likely to have experienced several updates/revisions between the time a request for the file was initiated and the time the end user received it. As a result, the end user may receive multiple versions of the same data file and have difficulty reconstructing it. A practical solution to this issue is to time-stamp each packet when a data file is input to the network by the GP node. Then, the end user is only allowed to reconstruct the file if he has collected a sufficient number of packets with the same time stamp.

8.4 Conclusion

In this chapter, we have presented an approach called M-FTP to improve the effectiveness of FTP applications in mobile opportunistic networks. Unlike traditional approaches, M-FTP does not need dedicated servers to form bridges between the Internet and mobile networks because it is a peer-to-peer application. It implements a *collaborative forwarding* feature to make better use of opportunistic connections among mobile peers and thereby improves the network capacity by exploiting the diversity of network mobility. Using simulations as well as real-world network scenarios, we evaluated M-FTP against a traditional approach called Mobile Hotspots, and the results demonstrate that our scheme can achieve better service ratios with moderate traffic overhead in all test cases. The effectiveness of the M-FTP scheme makes it an ideal solution that can facilitate mobile file transfer in mobile opportunistic networks.

References

[1] Apple, Inc., iPod + iTunes, http://www.apple.com/itunes/.

[2] Apple, Inc., QuickTime, Technologies, AAC Audio, http://www.apple.com/quicktime/technologies/aac/.

[3] CRAWDAD Project, http://crawdad.cs.dartmouth.edu/.

[4] Delay-Tolerant Network Simulator, http://www.dtnrg.org/code/dtnsim.tgz.

[5] UCSD Wireless Topology Discovery Project, http://sysnet.ucsd.edu/wtd/.

[6] L. Aalto, N. Gothlin, J. Korhonen, and T. Ojala. Bluetooth and WAP push based location-aware mobile advertising system. In *ACM International Conference on Mobile Systems, Applications and Services,* pages 49–58, 2004.

[7] L. Adamic and B. Huberman. Zipf's law and the Internet. *Glottometrics,* 3:143–150, 2002.

[8] E. Adar and B. A. Huberman. Free riding on Gnutella, http://www.firstmonday.org/issues/issue5_10/adar/, 2000.

[9] A. Bakre and B. R. Badrinath. Handoff and system support for indirect TCP/IP. In *The 2nd USENIX Symposium on Mobile and Location-Independent Computing,* pages 11–24, 1995.

[10] M. Balazinska and P. Castro. Characterizing mobility and network usage in a corporate wireless local-area network. In *ACM International Conference on Mobile Systems, Applications and Services,* pages 303–316, 2003.

[11] C. Buragohain, D. Agrawal, and S. Suri. A game theoretic framework for incentives in P2P systems. In *International Conference on Peer-to-Peer Computing,* 2003.

[12] J. Burgess, G. D. Bissias, M. Corner, and B. N. Levine. Surviving attacks on disruption-tolerant networks without authentication. In *ACM International Symposium on Mobile Ad Hoc Networking and Computing,* pages 61–70, 2007.

[13] J. Burgess, B. Gallagher, D. Jensen, and B. N. Levine. Maxprop: Routing for vehicle-based disruption-tolerant networks. In *IEEE Infocom,* pages 1688–1698, 2006.

[14] V. Bychkovsky, B. Hull, A. Miu, H. Balakrishnan, and S. Madden. A measurement study of vehicular internet access using in situ Wi-Fi networks. In *ACM International Conference on Mobile Computing and Networking,* pages 50–61, 2006.

[15] T. Camp, J. Boleng, and V. Davies. A survey of mobility models for ad hoc network research. *Wireless Communication and Mobile Computing Journal,* 2(5):483–502, 2002.

[16] A. Chaintreau, P. Hui, J. Crowcroft, C. D. Richard Gass, and J. Scott. Impact of human mobility on the design of opportunistic forwarding algorithms. In *IEEE International Conference on Computer Communications,* pages 1–13, 2006.

[17] L.-J. Chen, C.-L. Tseng, and C.-F. Chou. On using probabilistic forwarding to improve HEC-based data forwarding for opportunistic networks. In *IFIP Conference on Embedded and Ubiquitous Computing,* pages 101–112, 2007.

[18] L.-J. Chen, C.-H. Yu, T. Sun, Y.-C. Chen, and Hao-huaChu. A hybrid routing approach for opportunistic networks. In *ACM SIGCOMM Workshop on Challenged Networks,* pages 213–220, 2006.

[19] L.-J. Chen, C.-H. Yu, C.-L. Tseng, H. Hua Chu, and C.-F. Chou. A content-centric framework for effective data dissemination in opportunistic networks. *IEEE Journal of Selected Areas in Communications,* 26(5):761–772, June 2008.

[20] E. Chlebus and R. Ohri. Estimating parameters of the Pareto distribution by means of Zipf's law: Application to Internet research. In *IEEE Global Telecommunications Conference,* pages 1039–1043, 2005.

[21] M. C. Chuah and W.-B. Ma. Integrated buffer and route management in a DTN with message ferry. *Journal of Information Science and Engineering,* 23(4):1123–1139, July 2007.

[22] Y. Dalal and R. Metcalfe. Reverse path forwarding of broadcast packets. *Communications of the ACM,* 21:1040–1048, December 1978.

[23] S. Deering and R. Hinden. Internet protocol, version 6 (IPv6) specification. IETF RFC 2460, December 1998.

[24] K. E. Defrawy, M. E. Zarki, and G. Tsudik. Incentive-based cooperative and secure inter-personal networking. In *ACM International Workshop on Mobile Opportunistic Networks,* pages 57–61, 2007.

[25] H. Deng, W. Li, and D. P. Agrawal. Routing security in wireless ad hoc networks. *IEEE Communications Magazine,* 40:70–75, October 2002.

[26] M. Feldman, C. Papadimitriou, J. Chuang, and I. Stoica. Free-riding and whitewashing in peer-to-peer systems. In *ACM SIGCOMM Workshop on Practice and Theory of Incentives in Networked Systems,* pages 228–236, 2004.

[27] D. Goodman, J. Borras, N. Mandayam, and R. Yates. Infostations: A new system model for data and messaging services. In *IEEE Vehicular Technology Conference,* pages 969–973, 1997.

[28] M. Grossglauser and D. Tse. Mobility increases the capacity of ad-hoc wireless networks. *IEEE/ACM Transactions on Networking,* 10(4):477–486, August 2002.

[29] K. A. Harras, K. C. Almeroth, and E. M. Belding-Royer. Delay tolerant mobile networks (DTMNs): Controlled flooding in sparse mobile networks. In *IFIP Networking,* pages 1180–1192, 2005.

[30] D. Ho and S. Valaee. Mobile hot spot in railway systems. In *22nd Biennial Symposium on Communications,* 2004.

[31] X. Hong, M. Gerla, R. Bagrodia, and G. Pei. A group mobility model for ad hoc wireless networks. In *ACM International Workshop on Modeling, Analysis and Simulation of Wireless and Mobile Systems,* pages 53–60, 1999.

[32] D. Hughes, G. Coulson, and J. Walkerdine. Free riding on Gnutella revisited: The bell tolls? *IEEE Distributed Systems Online Journal,* 6(6):1, June 2005.

[33] P. Hui, A. Chaintreau, J. Scott, R. Gass, J. Crowcroft, and C. Diot. Pocket switched networks and human mobility in conference environments. In *ACM SIGCOMM Workshop on Delay Tolerant Networks,* pages 244–251, 2005.

[34] Z. Jiang and L. Kleinrock. An adaptive network prefetch scheme. *IEEE Journal on Selected Areas in Communications,* 16:358–368, April 1998.

[35] S. D. Kamvar, M. T. Schlosser, and H. Garcia-Molina. The Eigentrust algorithm for reputation management in P2P networks. In *International World Wide Web Conference,* pages 640–751, 2003.

[36] J. Kangasharju and A. Heinemann. Incentives for Opportunistic Networks. In *IEEE International Workshop on Opportunistic Networks,* pages 1684–1689, 2008.

[37] A. Krifa, C. Barakat, and T. Spyropoulos. Optimal buffer management policies for delay tolerant networks. In *IEEE Conference on Sensor and Ad Hoc Communications and Networks,* 2008.

[38] J. Leguay, T. Friedman, and V. Conan. DTN routing in a mobility pattern space. In *ACM SIGCOMM Workshop on Delay Tolerant Networks,* 2005.

[39] J. Leguay, T. Friedman, and V. Conan. Evaluating mobility pattern space routing for DTNs. In *IEEE International Conference on Computer Communications,* pages 1–10, 2006.

[40] Y. Liao, K. Tan, Z. Zhang, and L. Gao. Estimation based erasure-coding routing in delay tolerant networks. In *International Wireless Communications and Mobile Computing Conference,* 2006.

[41] A. Lindgren and A. Doria. Probabilistic routing protocol for intermittently connected networks. Technical report, draft-lindgren-dtnrg-prophet-01.txt, IETF Internet draft, July 2005.

[42] A. Lindgren and K. S. Phanse. Evaluation of queuing policies and forwarding strategies for routing in intermittently connected networks. In *IEEE International Conference on Communication System Software and Middleware,* page 1–10, 2006.

[43] D. A. Maltz and P. Bhagwat. MSOCKS: An architecture for transport layer mobility. In *IEEE Infocom,* pages 1037–1045, 1998.

[44] A. Matsumoto, M. Kozuka, K. Fujikawa, and Y. Okabe. TCP multi-home options, http://wit.jssst.or.jp/2003/papers/matsumoto.pdf.

[45] A. Nandan, S. Das, G. Pau, M. Gerla, and M. Y. Sanadidi. Co-operative download-ing in vehicular ad-hoc wireless networks. In *IEEE Conference on Wireless On-Demand Network Systems and Services(WONS),* 2005.

[46] A. Nandan, S. Das, B. Zhou, G. Pau, and MarioGerla. AdTorrent: Digital billboards for vehicular networks. In *IEEE/ACM International Workshop on Vehicle-to-Vehicle Communications,* 2005.

[47] J. Ott and D. Kutscher. Bundling the Web: HTTP over DTN. In *ICST Qshine Workshop on Networking in Public Transport,* 2006.

[48] V. N. Padmanabhan and J. C. Mogul. Using predictive prefetching to improve World Wide Web latency. *ACM SIGCOMM Computer Communication Review,* 26(3):22–36, July 1996.

[49] C. Perkins. IP mobility support for IPv4. IETF RFC 3344, August 2002.

[50] L. Ramaswamy and L. Liu. Free riding: A new challenge to peer-to-peer file sharing systems. In *36th Hawaii International Conference on System Sciences,* 2003.

[51] M. Schlager, B. Rathke, S. Bodenstein, and A. Wolisz. Advocating a remote socket architecture for Internet access using wireless LANs. *Journal of Mobile Networks and Applications,* 6(1):23–42, January/February 2001.

[52] T. Small and Z. J. Haas. The shared wireless infostation model—A new ad hoc networking paradigm (or where there is a whale, there is a way). In *ACM International Symposium on Mobile Ad Hoc Networking & Computing,* pages 233–244, 2003.

[53] T. Small and Z. J. Haas. Resource and performance tradeoffs in delay-tolerant wireless networks. In *ACM SIGCOMM Workshop on Delay Tolerant Networks,* 2005.

[54] A. Snoeren and H. Balakrishnan. An end-to-end approach to host mobility. In *ACM International Conference on Mobile Computing and Networking,* pages 155–166, 2000.

[55] T. Spangler. Push servers review. *PC Magazine,* pages 156–180, June 1997.

[56] R. Stewart, Q. Xie, K. Morneault, H. Schwarzbauer, T. T. I. Rytina, M. Kalla, L. Zhang, and V. Paxson. Stream control transmission protocol. IETF RFC 2960, October 2000.

[57] A. Vahdat and D. Becker. Epidemic routing for partially-connected ad hoc networks. Technical Report CS-2000-06, Duke University, 2000.

[58] V. Vishnumurthy, S. Chandrakumar, and E. G. Sirer. Karma: A secure economic framework for peer-to-peer resource sharing. In *Workshop on Economics of Peer-to-Peer Systems,* 2003.

[59] Y. Wang, S. Jain, M. Martonosi, and K. Fall. Erasure coding-based routing for opportunistic networks. In *ACM SIGCOMM Workshop on Delay Tolerant Networks,* pages 229–236, 2005.

[60] J. Widmer and J.-Y. L. Boudec. Network coding for efficient communication in extreme networks. In *ACM SIGCOMM Workshop on Delay Tolerant Networks,* 2005.

Chapter 9

Stationary Relay Nodes Deployment on Vehicular Opportunistic Networks

Joel J. P. C. Rodrigues
Instituto de Telecomunicações and University of Beira Interior

Vasco N. G. J. Soares
Instituto de Telecomunicações, University of Beira Interior, and Polytechnic Institute of Castelo Branco

Farid Farahmand
Sonoma State University

Contents

9.1 Introduction

The delay- and disruption-tolerant networking (DTN) architecture [1] is built as an overlay network, based on a store-carry-and-forward strategy, to provide communication facilities in challenged environments where sparse intermittent connectivity, high latency, asymmetric data rates, high loss rates, and even no end-to-end connectivity may exist.

In this type of network, a source node originates a bundle (message) and stores it until an appropriate communication opportunity becomes available. The bundle will be forwarded when the source node is in contact with an intermediate node that is closer (or in closer proximity) to the destination node. Afterward, the intermediate node stores the bundle and carries it while no new contact is available. This process is repeated and the bundle is relayed hop by hop until it eventually reaches its destination (Figure 9.1). Bundles have a finite time to live (TTL) and can be dropped if buffers in intermediate nodes overflow.

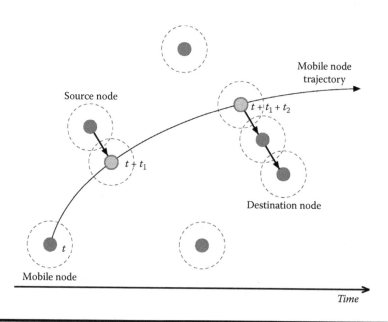

Figure 9.1   Store-carry-and-forward operation of DTN routing protocols.

An opportunistic network is a type of wireless mobile DTN where communication opportunities appear opportunistic, end-to-end paths may not exist, and intermittent connectivity is common. The concept of *opportunistic networking* has been widely applied on sparse and partitioned scenarios enabling nonreal time services and applications; for example, in Data MULEs [2], underwater networks [3], wildlife tracking sensor networks like ZebraNet [4], military ad hoc networks [5], people networks [6], and vehicular networks.

Data communication in vehicular networks is impacted by the high velocity of vehicles and their mobility patterns, which can cause rapid changes in the network topology. Environmental factors, such as physical obstacles or interferences, can limit the communication ranges. These factors, in conjunction with the node density and the distances/areas usually involved, result in a network with frequent fragmentation and intermittent communications, where there may not be an end-to-end path between every pair of nodes at all times. Message delivery routing approaches proposed for vehicular ad hoc networks are not appropriate for networks subject to episodic connectivity, frequent partitions, and connectivity disruptions [7]. Therefore, to cope with disconnection, vehicular opportunistic networks appear as a particular application of the DTN architecture concept together with its store-and-forward operation and routing/forwarding schemes.

Vehicular opportunistic networks have become a promising field of research into creating application scenarios such as the following: traffic condition monitoring, collision avoidance, emergency message dissemination, automatic tolling, free parking spot information, advertisements, and (for example) to gather data collected by vehicles, like road pavement defects [7,8,9,10]. Vehicles (e.g., cars, buses, and boats) can also be opportunistically exploited to offer a message relaying service by moving around a network and collecting messages from source nodes. A number of projects have been based on this general concept to propose transient networks to benefit developing communities and disaster recovery networks [11,12,13,14,15,16].

These different application scenarios raise a number of challenging issues, including network topology (known or not), node type (mobile, stationary), node design (energy constraints, storage capacity, physical link data rate, and transmission range), node mobility pattern (deterministic, stochastic, predictable), node cooperation, scheduling, traffic (static, dynamic), routing and forwarding protocols, buffer management schemes, and caching mechanisms.

Intermittent connected networks like vehicular opportunistic networks can benefit from the deployment of special nodes called *stationary relay nodes*. These nodes are fixed store-and-forward devices placed along vehicles' routes that exchange data with vehicles that pass the node. Therefore, they increase the number of connectivity opportunities between network nodes, thus improving the overall network performance. In this chapter, we first survey related work in the deployment of stationary relay nodes in vehicular opportunistic networks. We then present a study evaluating the effect of different numbers of relay nodes on the performance of four DTN routing protocols applied to vehicular opportunistic networks. We consider two scenarios (urban and rural) with distinct map areas, node density, and movement models. These results reveal the importance of stationary relay nodes to improve message delivery probability, even when no information is available about contact opportunities and traffic matrix.

The remainder of this chapter is organized as follows. Section 9.2 elaborates on the concept of the stationary relay node, reviewing the research efforts on using stationary relay nodes to improve the performance of vehicular opportunistic networks. Section 9.3 studies the performance impact of deploying stationary relay nodes on two different application scenarios for vehicular opportunistic networks. Finally, Section 9.4 concludes the chapter, providing a final summary of our study.

9.2 Stationary Relay Nodes

Mobility creates opportunities for vehicles to connect and communicate with one another. However, in vehicular opportunistic networks, intermittent connectivity is common and can be caused by various factors like high node mobility, low node density, energy constraints and power-management algorithms, short radio transmission ranges, or obstructed radio links. Networks can also be very sparse and are frequently partitioned (due to the large distances that may be involved). These conditions result in missed contact opportunities between vehicles and, consequently, decrease the message delivery ratio and increase message delivery delay. Eventually, it can produce resources contention in the network (e.g., buffers).

Vehicular opportunistic network overall performance can be improved by deploying additional stationary relay nodes in the network. *Stationary relay nodes* are fixed wireless nodes with store-and-forward capabilities that can be installed on road intersections, allowing passing vehicles to collect and leave data on them. Placing stationary relay nodes in strategic positions along vehicles' routes can potentially increase the number and frequency of contact opportunities. Thus, it will improve the overall network performance in terms of message delivery ratio and message delivery delay.

Figure 9.2 illustrates an example where a stationary relay node is deployed at a crossroad, creating an additional contact opportunity that did not exist before, since vehicles would not meet. When passing along the crossroad, vehicle *A* exchanges messages with the relay node at time *t*. Following a distinct trajectory, vehicle *B* passes along the relay node at a later time ($t + t_1$), collecting messages left there by vehicle *A*.

Subsection 9.2.1 reviews the literature concerning the use of stationary relay nodes to improve the overall performance of vehicular opportunistic networks.

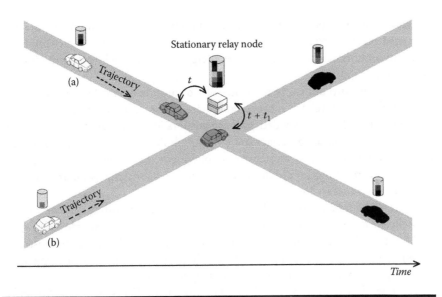

Figure 9.2 Vehicles exchanging data with a stationary relay node deployed in a road intersection.

9.2.1 Research Efforts on Increasing Vehicular Opportunistic Networks Performance with Stationary Relay Nodes

To the best of our knowledge, there is no (or very little) published work related to the problem of increasing the performance of vehicular opportunistic networks with the deployment of stationary relay nodes. In [17], authors use the term throwbox to refer to a stationary relay node, presenting a study that focuses on improving message delivery in DTNs using throwboxes. In the author's words, the work presents a "framework to systematically study the issues of throwbox deployment and routing." This framework considers distinct deployment scenarios differing on the information available about contact opportunities or traffic matrix. For each of these scenarios, different routing protocols are considered. The authors show that to maximize the throwboxes' effectiveness, their placement should be considered simultaneously with the routing algorithm.

The above work is extended in [18], where the authors conclude that, due to possible energy constraints, deploying throwboxes without using efficient power management schemes is minimally effective on the sparse connectivity in DTN scenarios. Therefore, they propose an energy-efficient hardware and software architecture for throwboxes, separating neighbor discovery from data transfer. Low-power, long-range, low-bit-rate radio hardware is used for neighbor discovery. Based on the information collected about neighbor discovery and mobility prediction, the high power, short-range, and high-bandwidth radio hardware will be turned on if necessary, but only during the duration of the contact opportunity. This architecture attempts to meet energy constraints while maximizing the packet delivery ratio.

In [19], the authors study the trade-offs of mobile networks enhanced with the deployment of relays, meshes, and wired base stations infrastructure. A cost–benefit analysis is provided, and the authors identify scenarios where wireless mesh or disconnected relays are a better choice than base stations or info stations, due to their ease and cost of deployment. However, a greater number of relays and mesh nodes are needed to achieve performance similar to that of base stations.

Markovian models are used to study the impact of adding throwboxes on the delivery delay [20] and resource consumption of Mobile Ad Hoc Networks (MANETs) [21]. These studies consider cases where throwboxes are fully disconnected or mesh connected, quantifying the impact of the number of throwboxes over the performance of routing protocols for each case.

A network architecture based on the concept of DTN, called a Vehicular Wireless Burst Switching Network (VWBS), is proposed in [15]. It suggests using *relay nodes* to provide connectivity between sparse regions, improving the overall network performance in terms of average packet delay and packet loss. The work proposes several heuristic algorithms to provide a solution to the *relay node placement problem*. This problem is formulated as follows: Given a previously known network topology, vehicle mobility model, traffic matrix, and the number of available relay nodes, the goal is to determine where relay nodes should be placed at road intersections in order to maximize the network performance in terms of network cost (minimizing relay node), delay (minimizing delivery time), or both (minimizing relay node and delivery time).

This previous work is extended in [22], where the authors show that the problem of optimal relay node placement is a nondeterministic polynomial-time hard (NP-hard) problem and propose an integer linear programming (ILP) formulation for this problem. Two heuristic algorithms are also presented. One of the algorithms aims at minimizing the number of hop counts between source–destination terminal nodes, while the other aims at minimizing the average message delivery time to the destination. Nevertheless, both algorithms also attempt to minimize the number of required relay nodes in the network.

In [23], the authors evaluate the influence of relay nodes in a vehicular delay-tolerant network applied to a limited city area. They present a simulation-based study evaluating the impact of varying the number of relay nodes in scenarios with different numbers of vehicles. An opportunistic environment was assumed without knowledge of the traffic matrix. The authors concluded that relay nodes significantly improve the message delivery ratio in network environments with different node densities and are prone to frequent connectivity disruptions.

9.3 Impact of Stationary Relay Nodes on the Performance of Vehicular Opportunistic Networks

To analyze the impact of stationary relay nodes on the performance of vehicular opportunistic networks, a simulation study is performed using the Opportunistic Network Environment (ONE) Simulator [24]. The next subsections will present two distinct application scenarios for vehicular opportunistic networks. First, we evaluate a city scenario where data is exchanged between vehicles moving in a small area (Subsection 9.3.1). Subsection 9.3.2 presents a dispersed vast rural region, where vehicles are used as the communication infrastructure for the network in order to provide asynchronous Internet access to undeveloped remote areas.

For each of those scenarios, we analyze the number of contact opportunities registered per hour between network nodes when different numbers of stationary relay nodes are deployed in predefined map locations. We examine the impact of stationary relay nodes on the performance of the widely applicable multiple-copy DTN routing strategies Epidemic [25], MaxProp [26], PRoPHET [27], and Spray and Wait [28].

Epidemic is a flooding-based routing protocol where nodes exchange the messages they do not have. In an environment with infinite buffer space and bandwidth, this protocol performs better than the others in terms of message delivery ratio and latency, providing an optimal solution. MaxProp prioritizes the schedule of messages transmitted to other nodes and also the schedule of messages to be dropped.

PRoPHET is a probabilistic routing protocol that considers a history of encounters and transitivity. It considers that nodes move in a nonrandom pattern and applies *probabilistic routing*. The Spray and Wait protocol creates a number of copies N to be transmitted (*sprayed*) per message. In its normal mode, a source node A forwards the N copies to the first M different nodes encountered. In the binary mode, any node A that has more than one message copies and encounters any other node B that does not have a copy, forwards to B the number of $N/2$ message copies, and keeps the rest of the messages. A node with one copy left only forwards it to the final destination.

The network performance is measured according to two metrics: the overall message delivery ratio (measured as the relation of the number of unique delivered messages to the number of messages sent) and the message delivery delay (measured as the time between message creation and corresponding delivery). Although other metrics are essential to assess the routing protocols' performance (e.g., hop count, routing overhead), given the focus of our work, we selected the metrics considered to be the most important indicators of the gains introduced by stationary relay nodes.

Both scenarios assume a fully cooperative opportunistic environment without knowledge of the traffic matrix and contact opportunities. We simulate the creation and message exchange during a period of 12 hours (e.g., from 08:00 to 20:00), and measure the performance differences when stationary relay nodes are added to the network. In order to make the simulation results statistically credible, a large

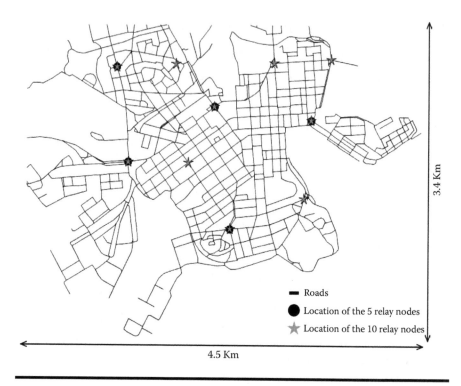

Figure 9.3 Helsinki simulation area with the locations of the stationary relay nodes.

number of simulation runs with different random number generation seeds were carried out.

9.3.1 Scenario I: Urban Connectivity

The first scenario considers the use of a vehicular opportunistic network to enable connectivity in a city, allowing the use of non-real-time applications, such as notification of blocked roads or advertisements. For the simulation scenario, we use a map-based model of a small part of Helsinki, Finland (Figure 9.3).

9.3.1.1 Simulation Settings

This scenario assumes the simulation of 20 vehicles moving between random map locations. Once a vehicle reaches a destination, it randomly waits from 5 to 15 minutes. Then it selects a new random map location and a random speed between 10 and 50 km/h. The vehicle moves to the new destination using the shortest available path. Each vehicle has a 150-Mbyte FIFO (first-in first-out) message buffer. Messages are exchanged between random source and destination vehicles. It used an intermessage creation interval in the range [0, 15] (seconds) of uniformly distributed random values. Message size is in the range [500 KB, 2 MB] of uniformly distributed random values. All the messages exchanged in the simulations have an infinite time to live (TTL).

We are interested in studying the impact of using 0, 5, and 10 stationary relay nodes, each with a 500-Mbyte FIFO message buffer size. Stationary relay nodes are placed at the predefined map locations shown in Figure 9.3. Their coverage cells do not intersect, so they are not able to communicate directly with each other, only with the vehicles. In addition, vehicles exchange data with each other.

Network nodes connect to each other using IEEE 802.11b with a data rate of 6 Mbit/s (the 802.11b approximate throughput according to [29]), and a transmission range of 50 meters using omnidirectional antennas.

9.3.1.2 Results Analysis

As expected, the analysis of Figure 9.4 shows that deploying stationary relay nodes increases the number of contact opportunities per hour between all network nodes. Introducing 10 stationary relay nodes increases the number of contacts at a rate of roughly a factor of two per hour. This effect suggests that relay nodes will contribute to increasing the number of messages exchanged between vehicles.

Figure 9.5 shows the message delivery probability for each considered routing protocol, assuming 0, 5, and 10 stationary relay nodes. This figure indicates that when no stationary relay nodes are deployed in the network, MaxProp performs better than the other protocols in terms of message delivery probability. The other routing protocols register similar performance. When relay nodes are introduced on

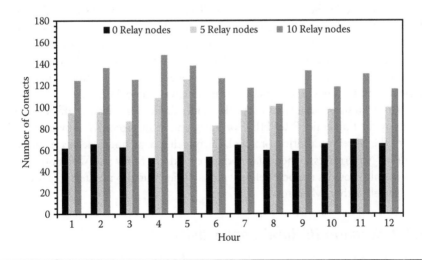

Figure 9.4   Number of contacts per hour between all network nodes.

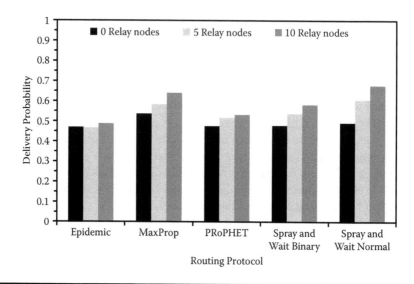

Figure 9.5   Message delivery probability.

the network scenario, all routing protocols increase their delivery probabilities except Epidemic. This behavior is due to its poor utilization of the network resources.

When 5 stationary relay nodes are deployed, MaxProp increases the delivery probability by 5%, PRoPHET improves it by 4%, Spray and Wait binary variant (with 12 message copies) increases it by 6%, and finally, Spray and Wait normal variant (with the same number of message copies) augments delivery probability by 11%. Adding more stationary relay nodes to the network (10 relay nodes) causes a further increase in the delivery ratio. In effect, MaxProp and PRoPHET increase their message delivery probability by 5% and 2%, respectively. Spray and Wait binary variant augments by 5%, whereas the normal variant has a gain of 7%.

Furthermore, it can be observed that Spray and Wait (normal variant) is the routing protocol that experiences a higher benefit from the introduction of the stationary relay nodes, and registers the best delivery probabilities.

Figure 9.6 shows that stationary relay nodes do not significantly affect the average delay for Epidemic, PRoPHET, or Spray and Wait protocols. However, in the case of the MaxProp routing protocol, introducing relay nodes decreases the average delay. Nevertheless, when relay nodes are used, the Spray and Wait normal variant not only provides the best delivery probabilities, but also registers much smaller average delays than MaxProp.

9.3.2 Scenario II: Rural Connectivity

In this second scenario, we consider the use of a vehicular opportunistic network to provide low-cost asynchronous Internet access on a vast geographical area with a

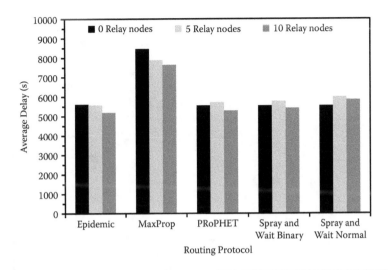

Figure 9.6   Average message delay.

sparse population density and without a network infrastructure. In [15,30] is proposed an architecture model for such networks based on the following node types: terminal nodes, mobile nodes, and relay nodes. Terminal nodes (*traffic sources*) are located in isolated regions and provide connections for end users. One of the terminal nodes has an Internet access (*traffic sink*). Mobile nodes (i.e., vehicles) are opportunistically exploited to carry data between the traffic sources and the traffic sink.

For this simulation scenario we use a map-based model representation of the *Serra da Estrela Region*, a Portuguese mountain region that covers an area of approximately 2500 Km². This map-based model is shown on Figure 9.7.

9.3.2.1 Simulation Settings

We select 24 village locations to place terminal nodes that act as *traffic sources* (Figure 9.4). Each traffic source has a 2-Gbyte FIFO message buffer, and generates messages using an intermessage creation interval in the range [15, 30] minutes with uniformly distributed random values. Each message has a size in the range [500 KB, 2 MB] with uniformly distributed random values. All the messages exchanged in the simulations have an infinite time to live (TTL). Their destination address is the terminal node connected to the Internet that acts as the traffic sink.

We have a group of 12 vehicles moving on roads, each one with a 2-Gbyte FIFO message buffer. When a vehicle reaches a terminal node, it randomly waits from 30 to 60 minutes. Then, it selects its next destination in accordance with a probability, assuming a 90% probability of selecting a random traffic source as its next destination and a 10% probability of selecting the traffic sink. A random speed between 30 and 80 km/h is selected, and the vehicle moves there using the shortest path.

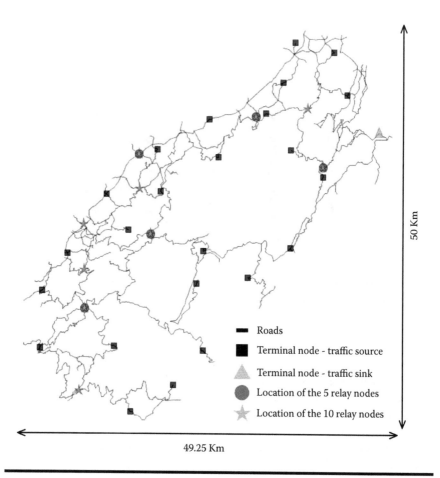

Figure 9.7 *Serra da Estrela Region* **simulation area with the locations of the terminal nodes and the stationary relay nodes.**

We are interested in the impact of using 0, 5, or 10 stationary relay nodes, with a 2-Gbyte FIFO message buffer each, located as shown in Figure 9.7. Network nodes connect to each other using 802.11b, with a data rate of 6 Mbit/s and a transmission range of 350 meters using omnidirectional antennas. Terminal nodes and relay nodes exchange data only with vehicles. Vehicles can communicate with each other.

9.3.2.2 Results Analysis

The large size of this second scenario, the low network node density, and the above-mentioned vehicle mobility model, lead to a small number of contacts registered per hour. Deploying stationary relay nodes in this scenario is a complex task, since we assumed that we do not have any information about the transmission

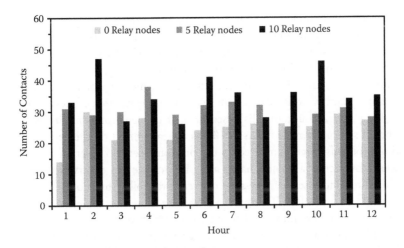

Figure 9.8   Number of contacts per hour between all network nodes.

opportunities and traffic matrix. Similar to the previous scenario, we used a non-uniform strategy [19] to place relay nodes, positioning them preferentially in the core network. As may be seen in Figure 9.8, this simple strategy produces significantly positive results even in this sparse scenario, increasing the number of contact opportunities between the network nodes.

Figure 9.9 shows the message delivery probability for each considered routing protocol, assuming 0, 5, and 10 stationary relay nodes. As seen in this figure, when

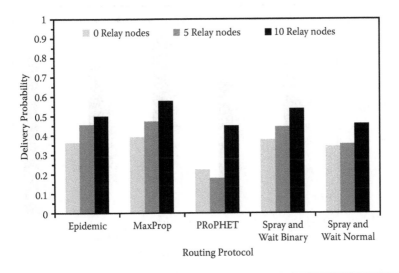

Figure 9.9   Message delivery probability.

no stationary relay nodes are deployed in the network, MaxProp performs better than the other protocols in terms of message delivery probability. It may be observed that deploying 5 stationary relay nodes provides up to a 9% gain in message delivery probability for the Epidemic routing protocol, an 8% gain for MaxProp, a 7% gain for the Spray and Wait binary variant (with 36 message copies), and a 1% gain for the Spray and Wait normal variant (with the same number of message copies).

Taking into account the dispersed area of this environment, increasing the number of stationary relay nodes to 10 has a greater effect on the delivery ratio of routing protocols when compared to the previous scenario (Section 9.3.1), where vehicles move across a much smaller area. In fact, we can observe that Epidemic and MaxProp increase their message delivery probability by about 5% and 10%, respectively, and PRoPHET improves 27%. The same is observed with the Spray and Wait variants that improve 9% and 10%, respectively.

In this scenario, the network nodes' large buffers attenuate Epidemic's poor utilization of the network's resources. PRoPHET's probabilistic routing approach registers the lowest delivery probabilities when 0, 5, or 10 relay nodes are used. Finally, for this scenario, it can also be concluded that MaxProp performs better than the other protocols, and that the Spray and Wait binary variant with message replication technique is more efficient that the normal variant.

As may be observed in Figure 9.9, PRoPHET not only has the lowest delivery probabilities, but also the worst performance with respect to message average delay (shown in Figure 9.10). The remaining routing protocols register similar values for this performance metric across all simulations. The small increase on the average

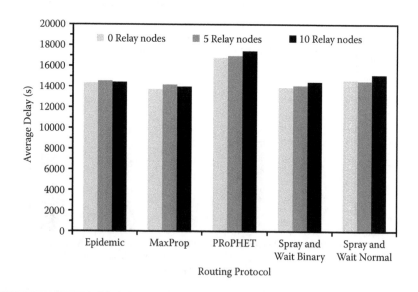

Figure 9.10 Average message delay.

delay when relay nodes are introduced is related to the time that messages spend in their buffers, waiting for a vehicle to pick up and deliver them to the traffic sink.

Based on results presented in Figure 9.9 and Figure 9.10, we can also conclude that by deploying stationary relay nodes, more messages are successfully delivered to the traffic sink without significantly increasing the time needed to deliver them. The large delivery delays registered in this scenario are due to the vast geographical area.

9.4 Conclusions

Vehicular opportunistic networking has become a hot research topic that has been rapidly growing in the last few years. In this chapter, we began by presenting a brief overview of delay-tolerant and opportunistic network concepts that serve as a basis for vehicular opportunistic networks.

Overall, vehicular opportunistic networking is a challenging topic with many open research issues that require innovative approaches. We focused our study on the research problem related to stationary relay node deployment to improve the performance of these networks. Stationary relay nodes have been proposed to cope with the challenges caused by the following circumstances: high mobility of nodes, low node density, intermittent connectivity, and sparse and partitioned networks. A comprehensive literature review proposed in this area was presented.

Simulations were conducted to evaluate the impact of adding stationary relay nodes in the performance of four DTN routing protocols applied to vehicular opportunistic networks. For each simulated environment, different application scenarios, map areas, node density, and vehicle movement models were combined.

Our results confirm that stationary relay nodes have a very positive impact on the performance of the routing protocols considered in this study, both in terms of message delivery probability and message delivery delay. These results also corroborate their influence on the number of opportunistic contacts (transmission opportunities) that occur between network nodes.

In this study, a cooperative opportunistic environment without knowledge of the traffic matrix and contact opportunities was assumed. Therefore, relay nodes were placed at the core network. If information about contact opportunities and the traffic matrix were previously available, the relay node placement could be considered simultaneously with routing, and the overall network capacity would be enhanced [17].

Acknowledgments

Part of this work has been supported by the *Instituto de Telecomunicações*, Next Generation Networks and Applications Group (NetGNA), Portugal, in the framework of the Project VDTN@Lab, and by the Euro-NF Network of Excellence from the Seventh Framework Programme of the European Union.

References

[1] V. Cerf, S. Burleigh, A. Hooke, et al. Delay-tolerant networking architecture, ftp://ftp. rfc-editor.org/in-notes/rfc4838.txt, April 2007.

[2] S. Jain, R. Shah, W. Brunette, G. Borriello, and S. Roy. Exploiting mobility for energy efficient data collection in wireless sensor networks. *ACM/Kluwer Mobile Networks and Applications (MONET)* 11(3):327–339, 2006.

[3] J. Partan, J. Kurose, and B. N. Levine. A survey of practical issues in underwater networks. Paper presented at *1st ACM International Workshop on Underwater Networks*, in conjunction with ACM MobiCom 2006, Los Angeles, California, September 25, 2006.

[4] P. Juang, H. Oki, Y. Wang, et al. Energy-Efficient Computing for Wildlife Tracking: Design Tradeoffs and Early Experiences with ZebraNet. *ACM SIGOPS Operating Systems Review* 36(5):96–107, 2002.

[5] J. Jormakka, H. Jormakka, and J. Väre. A lightweight management system for a military ad hoc network. In *Lecture Notes in Computer Science*. Berlin/Heidelberg: Springer, 2008.

[6] N. Glance, D. Snowdon, and J.-L. Meunier. Pollen: Using people as a communication medium. *Computer Networks: The International Journal of Computer and Telecommunications Networking* 35(4):429–442, 2001.

[7] L. Franck and F. Gil-Castineira. Using delay-tolerant networks for car2car communications. Paper presented at *IEEE International Symposium on Industrial Electronics 2007 (ISIE 2007)*, Vigo, Spain, June 4–7, 2007.

[8] R. Tatchikou, S. Biswas, and F. Dion. Cooperative vehicle collision avoidance using inter-vehicle packet forwarding. Paper presented at IEEE Global Telecommunications Conference (IEEE GLOBECOM 2005), St. Louis, Missouri, November 28–December 2, 2005.

[9] I. Leontiadis and C. Mascolo. GeOpps: Geographical opportunistic routing for vehicular networks. Paper presented at *IEEE International Symposium on a World of Wireless, Mobile and Multimedia Networks 2007 (WoWMoM 2007)*, Espoo, Finland, June 18–21, 2007.

[10] O. Brickley, C. Shen, M. Klepal, A. Tabatabaei, and D. Pesch. A data dissemination strategy for cooperative vehicular systems. Paper presented at *Vehicular Technology Conference 2007 (VTC2007)*, at Dublin, Ireland, 2007.

[11] W. Zhao, M. Ammar, and E. Zegura. A message ferrying approach for data delivery in sparse mobile ad hoc networks. Paper presented at *The 5th ACM International Symposium on Mobile Ad Hoc Networking and Computing (MobiHoc 2004)*, Tokyo, Japan, May 24–26, 2004.

[12] W. Zhao and M. H. Ammar. Message ferrying: Proactive routing in highly-partitioned wireless ad hoc networks. Paper presented at *The 9th IEEE Workshop on Future Trends of Distributed Computing Systems*, San Juan, Puerto Rico, May 28–30, 2003.

[13] A. Pentland, R. Fletcher, and A. Hasson. DakNet: Rethinking connectivity in developing nations. *IEEE Computer*, 78–83, January 2004.

[14] Wizzy Digital Courier. *Wizzy Digital Courier—Leveraging locality*, http://www.wizzy. org.za/, accessed January 2008.

[15] F. Farahmand, A. N. Patel, J. P. Jue, V. G. Soares, and J. J. Rodrigues. Vehicular wireless burst switching network: Enhancing rural connectivity. Paper presented at *The 3rd IEEE Workshop on Automotive Networking and Applications (Autonet 2008)*, co-located with IEEE GLOBECOM 2008, New Orleans, Louisiana, December 4, 2008.

[16] N4C and eINCLUSION. *Networking for communications challenged communities: Architecture, test beds and innovative alliances*, 2009, http://www.n4c.eu/, accessed January 2009.

[17] W. Zhao, Y. Chen, M. Ammar, et al. Capacity enhancement using throwboxes in DTNs. Paper presented at *IEEE International Conference on Mobile Adhoc and Sensor Systems (MASS)*, 2006.

[18] N. Banerjee, M. D. Corner, and B. N. Levine. An energy-efficient architecture for DTN throwboxes. Paper presented at *INFOCOM 2007, 26th IEEE International Conference on Computer Communications*, 2007.

[19] N. Banerjee, M. D. Corner, D. Towsley, and B. N. Levine. Relays, base stations, and meshes: Enhancing mobile networks with infrastructure. Paper presented at *The 14th ACM International Conference on Mobile Computing and Networking (ACM MobiCom)*, San Francisco, California, September 2008.

[20] M. Ibrahim, A. A. Hanbali, and P. Nain. Delay and resource analysis in MANETs in presence of throwboxes. *Performance Evaluation*, 64(9–12):933–947, 2007.

[21] S. Corson and J. Macker. Mobile ad hoc networking (MANET): Routing protocol performance issues and evaluation considerations, January 1999, http://www.ietf.org/rfc/rfc2501.txt.

[22] Farahmand, I. Cerutti, A. N. Patel, Q. Zhang, and J. P. Jue. Relay node placement in vehicular delay-tolerant networks. Paper presented at *IEEE Global Telecommunications Conference 2008 (IEEE GLOBECOM 2008)*, New Orleans, Louisiana, November 30–December 4, 2008.

[23] V. N. G. J. Soares, F. Farahmand, and J. J. P. C. Rodrigues. A layered architecture for vehicular delay-tolerant networks. Paper presented at *The 14th IEEE Symposium on Computers and Communications (ISCC'09)*, Sousse, Tunisia, July 5–8, 2009.

[24] A. Keränen, J. Ott, and T. Kärkkäinen. The ONE simulator for DTN protocol evaluation. Paper presented at *SIMUTools'09: 2nd International Conference on Simulation Tools and Techniques*, Rome, Italy, March 2–6, 2009.

[25] A. Vahdat and D. Becker. 2000. Epidemic routing for partially-connected ad hoc networks. Duke University.

[26] J. Burgess, B. Gallagher, D. Jensen, and B. Levine. 2006. MaxProp: Routing for vehicle-based disruption-tolerant networks. Paper presented at *INFOCOM 2006, The 25th IEEE International Conference on Computer Communications*, Barcelona, Spain, April 23–29, 2006.

[27] A. Lindgren and A. Doria. Probabilistic routing protocol for intermittently connected networks, November 17, 2008, http://www.ietf.org/internet-drafts/draft-irtf-dtnrg-prophet-01.txt.

[28] T. Spyropoulos, K. Psounis, and C. S. Raghavendra. Spray and wait: An efficient routing scheme for intermittently connected mobile networks. Paper presented at *ACM SIGCOMM 2005, Workshop on Delay-Tolerant Networking and Related Networks (WDTN-05)*, Philadelphia, Pennsylvania August 22–26, 2005.

[29] Cisco Systems, Inc. *Capacity Coverage & Deployment Considerations for IEEE 802.11g*, 2005, http://www.cisco.com/en/US/products/hw/wireless/ps4570/products_white_paper09186a00801d61a3.shtml, accessed December 2008.

[30] V. N. G. J. Soares, F. Farahmand, and J. J. P. C. Rodrigues. Improving vehicular delay-tolerant network performance with relay nodes. Paper presented at *5th Euro-NGI Conference on Next Generation Internet Networks (NGI 2009)*, Aveiro, Portugal, July 1–3, 2009.

Chapter 10

Connection Enhancement for Mobile Opportunistic Networks

Weihuang Fu, Kuheli Louha, and Dharma P. Agrawal
University of Cincinnati

Contents

10.1 Introduction

Unlike most of the existing networks, connection of a mobile node (MN) in a mobile opportunistic network (MON) does not always exist. An MN may connect to another MN in a MON for some duration and disconnect afterward. A new connection to the MN in question may be formed at a later time or connections it already had in the past with some MNs may be reinstated. As a result, the routing path from a source MN to a destination MN may not exist at a given time. However, from a longtime routing point of view, the path could be available in the network if we overlap intermittent connections in different time frames, which provides an opportunity for a message to be delivered from a source to a destination. A new trend is emerging with network technology and theory is being developed to support message delivery in such a network.

In the immediate past, a similar network research concept is the delay-tolerant/disruption-tolerant network (DTN) [1], which is used for applications that allow longer delays and disruptions of the messages being forwarded. The intention of DTN is to interconnect multiple regional networks, where the nodes moving from one regional network to another is used to help deliver the messages. The main features of DTNs are the heterogeneity of regional networks, storage and forwarding of messages, overlay protocol for internetwork end-to-end data transfer, and so on.

Although the concepts of DTNs and MONs widely overlap, they have some distinguishing characteristics. The MON concept is more general and focused on opportunistic connections and routing, while DTNs emphasize the delay- and disruption-tolerant attribute of the application and interconnection of regional networks. There are only a few nodes in a DTN acting as the gateway and equipped with heterogeneous interfaces for internetwork connectivity. A MON assumes that each MN is equipped with multiple heterogeneous interfaces and is able to interconnect multiple networks at the same time. In addition, the opportunistic

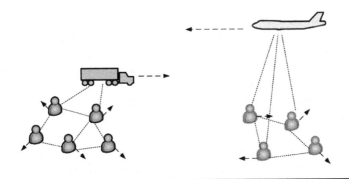

Figure 10.1   Example of the MON architecture.

connection of a DTN may be possible only to a limited number of nodes located at the boundary of a regional network, while the nodes within the regional network are well connected. In contrast, a MON assumes that each MN has mobility and that connections between MNs are intermittent.

An obvious example of a MON is tracing sensors installed on wild animals [2,3], where connections between wireless sensors and data collectors can be established and messages can be delivered when animals pass by data collectors installed at some fixed locations. It is obvious that when animals move frequently, they have a better opportunity to connect and deliver messages. Communication between sensor and data collector is a single-hop wireless connection in the above case. If we allow messages to be exchanged between sensors on different groups of animals when they are within each other's communication region, the message delivery probability could be enhanced by leveraging multiple hop transmissions.

Figure 10.1 shows a general case of MON network architecture. One group of people is on the left and the other on the right. A truck moves from the left to the right and an airplane flies from the right to the left. These are moving objects and mobile devices carried on them can communicate with others within their transmission range if they are equipped with the same type of wireless interfaces and are used on the same protocols. A mobile device with its carrier is called as an MN in a MON. An MN may not always connect to other MNs in the MON. The time from creating a connection to a break between two MNs within a transmission region is defined as *connection duration*. Intuitively, higher mobility will reduce the connection duration while it increases the number of connections for MNs. For example, if the truck in Figure 10.1 moves faster, the connection duration will be less for each MN in its communication region, while the number of connections in a unit time, called *connection frequency*, could be increased. Some of the MNs may be equipped with multiple heterogeneous interfaces that are able to access different types of networks. As shown in Figure 10.1, the connection between the human being and the truck relies on whether they have the same type of wireless interface. The same is applicable to the connections between the airplane and the group of people on the right. If not every MN in the network is equipped with the same interface, a heterogeneous MN is used as the bridge or gateway for message forwarding. Since an MN could have multiple interfaces, it can enable multiple connections to multiple networks at the same time, thereby increasing the connection duration and connection frequency in the MON.

From the definition and architecture of a MON, we can identify the following features.

- The connection between any two MNs is opportunistic and intermittent. An MN needs to scan the network or the channel periodically to discover possible connections to other MNs or the infrastructure. Beacon signals are required and frequently used for this purpose.
- Due to dynamic changes in the network topology and a wider network scope, it is impossible for every MN to have information about a whole network topology. Only one- or two-hop network connectivity information may be available to MNs.
- Due to node mobility, a connection may not last for too long. For completeness of any transmission, an MN may require an acknowledgment or estimate of the connection duration. It is also helpful if an MN can estimate the interconnection duration (i.e., the time interval between two connections).
- Message forwarding is based on a store-carry-forward scheme, as the connection may not be available all the time. A large space is required to store received messages before the MN has a chance to connect. Existing network protocol architectures are not resilient to disruption in communication links.
- An MN may be equipped with multiple interfaces and can connect to different networks. An MN equipped with multiple heterogeneous interfaces can forward messages in one or more networks and have a better chance to deliver messages. So, there may be an issue of how to select the networks.
- In a MON, a physical connection between MNs is intermittent. Opportunistic access to a single network may not provide satisfactory quality of service (QoS) due to frequent disconnections.
- For message forwarding, it needs to address how to select the next-hop MN. The routing path may not be there at that time because the knowledge of the network topology is limited.
- An MN may exploit a successful message delivery opportunity by duplicating a message to different next-hop MNs. Multiple copies along different paths can benefit from path diversity, but this also introduces a new problem with respect to how many copies are needed and which set of MNs should be chosen for forwarding.

To deal with these features, new protocols have to be developed, and there are still many challenging open issues. Although there are many difficulties in developing a MON, underlying unique characteristics still attract researchers' attention as a step toward the next-generation Internet. First, MON technology can potentially interconnect existing networks to form a larger universal network. Second, MON can provide higher throughput [4] than the upper bound of the current wireless network capacity due to its unique message transfer pattern. In addition, it is a low-cost deployment and mobility of MNs adds to the savings relative to investment. For some applications, such as delay- and disruption-tolerant applications, they do not need connectivity all the time.

This chapter considers strategies for enhancing connections and provides insight on the state-of-the-art research on connections in mobile and heterogeneous networks, message forwarding, and the network selection process in MONs. The connection models in mobile and heterogeneous networks form the foundation of MONs. Mobility models indicate the connection duration and its frequency for a given mobility pattern, network size, and so on. In contrast to a conventional network, MNs carry messages in MONs and move around in the network. So, the mobility model also needs to determine how far an MN can travel while carrying a message. It potentially has a positive impact on the network capacity. Heterogeneous connection enables MNs to have better opportunities for having connections to a heterogeneous network. Both mobility and heterogeneous connectivity make the problem extremely complex and there is a need for creating a mathematical tool for network selection and the associated message forwarding. When an MN has several interfaces, there may be multiple heterogeneous networks present in the area. The connection selection involves a series of procedures taking network conditions and handover steps into account.

Message forwarding entails how many messages are to be forwarded and to which MNs. The strategies are mainly divided into two categories: single-copy and multicopy forwarding. Single-copy forwarding works in a conventional zigzag path and requires message storage while the connection is not available. Multicopy forwarding explores multiple paths to enhance message delivery probability and reduce associated delivery latency. However, this is at the cost of storage space at various relay MNs. Since an MN will have multiple connections to different MNs while moving, it is a challenging problem to decide which one it should forward the message to. It is unrealistic to assume that the MN has the complete network information. The network condition changes very often and a routing path from a source to a destination may not always exist. The forwarding decision has to be on a per-hop basis (i.e., a per-connection basis).

The rest of this chapter is organized as follows: The impact of mobility and heterogeneous connection on the performance of the MON are investigated in Section 10.2. Section 10.3 considers various network selection approaches, and Section 10.4 investigates numerous message forwarding approaches. Finally, the chapter is concluded in Section 10.5.

10.2 Mobility and Heterogeneous Connections

The performance of message delivery in MONs depends on two essential aspects: mobility and heterogeneous connections. Mobility enables MNs to enhance the opportunity to access different MNs while they are within the transmission range. However, for a pair of MNs within the transmission distance, creation of a successful connection also requires that they have the same type of network interface. An MN with multiple heterogeneous interfaces will have better communication opportunities with the ability to connect to different networks.

Benefiting from mobility, MNs without any previous connections can move to a region that could enable connecting or forwarding messages to other MNs. Of course, it may get disconnected at a future time if one of the MNs moves out of the transmission range. From a long duration point of view, connection between two MNs may be present with certain probability. As shown in Figure 10.2, the connection status of an MN can be defined by two states: connected (state C) and disconnected (state D). When it is in state C, the probability that the connection is broken at the next time slot is b, and the probability that the connection remains connected is $1 - b$. Similarly, when it is in state D, the probability that the connection is created at the next time slot is g, and the probability that the MN remains disconnected is $1 - g$. The values of the parameters highly depend on the mobility attributes of MNs in the network.

MNs may move with different speeds: nomadic walk, vehicular speed, very high vehicular speed, and so on. When the speed is high, the frequency of connections between two MNs within a communication region (i.e., the number of connections) increases accordingly. As indicated by the state diagram in Figure 10.2, g increases if mobility is high. However, the connection duration is decreased if MNs have high mobility in the network, which is indicated by the increase in b. Thus, higher mobility has both positive and negative effects on the network performance. The pattern of mobility may also lead to geographic nonuniform distribution of connection duration and frequency, which means that different MNs do not have identical parameters for their state transition. Taking an example of a MON

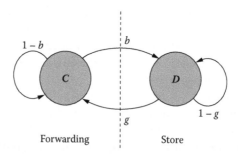

Figure 10.2　Connection state transit diagram.

on a highway, the connectivity between vehicles moving in the same direction will be relatively stable, while the connection duration is very short for the vehicles moving in the opposite direction.

The network connectivity is also greatly affected by the density of MNs. Higher node density not only increases the number of connections, but also increases the duration of the connection. As a result, it can effectively reduce the message delivery latency and increase the delivery probability. MNs with multiple heterogeneous interfaces can further enhance the chances of connections to other nodes. While parameter g in the state diagram of Figure 10.2 increases, the duration of state D becomes shorter. It is worth noting that the connection can be switched from one MN to another with the same interface type or a heterogeneous type. Handover can be performed during such a switching process. We consider this as connection state C remaining the same, as there is no interruption of connection from the point of view of an upper layer.

It has been proved [5] that a single copy of a message may not reach the destination with high probability if the node density is not adequate above a given threshold. The density threshold is primarily related to the specified mobility model. The node density of a network may be sufficient or not, and MNs may have to utilize a message duplication method so as to enhance the chances of achieving a finite delay. Although message duplication increases the message delivery probability, it also quickly decreases the network capacity.

In Figure 10.2, state D has a negative impact on the network performance. MNs cannot perform transmission during the disconnected state, and storage space is needed to buffer the messages. However, a mobile MN is able to move messages from one location to another, which is equivalent to wireless transmission by the same distance. So, the movement during the disconnected state also indirectly contributes to the network transmission.

10.2.1 Mobility Models

To investigate these problems, an accurate mobility model of MNs is necessary so that it can serve as an accurate mathematical tool in describing statistical attributes of MNs with speed and direction in a defined geographic region. There are many different mobility models with respect to changes in speed and direction. Without loss of generality, we consider the mobility models in a two-dimensional region.

The ***random walk mobility model*** [6] is used very frequently for mobile networks. For an MN in a network, velocity vector $\vec{v} = \langle v, \theta \rangle$ is determined by the speed v and the angle θ. It changes every time MN moves for time t or distance d. When the velocity changes, the speed is arbitrarily selected between predefined minimum and maximum speeds, $[v_{min}, v_{max}]$, and the direction is uniformly selected from $[0, 2\pi)$. The movement of an MN is primarily indicated by two variables: speed and direction. From the definition of the random walk mobility model, it is a memoryless mobility process. The change in speed and direction has no dependence on any previous MNs' mobility. It may be slightly different from a real application. The movement of an MN in a real application may not be completely memoryless. Nevertheless, the model still retains many essential attributes of a mobile network and is widely used by researchers.

The ***random waypoint mobility model*** [7] introduces a concept of pause time τ during movement. An MN stops at a location for time τ before it starts moving again. When the MN moves, it randomly selects a destination point in the defined region and moves toward the destination with a speed randomly selected between $[v_{min}, v_{max}]$. The random waypoint mobility model is also a memoryless process. The movement of nodes only depends on three variables: pause time, speed, and the next destination location. It is worth noting that the introduced pause time may help increase connection duration. This model is more likely to represent the movement and actions of a human walk.

The ***city mobility model*** [8] is a more realistic model for the MNs in an urban area. Unlike the models described above, movement speed and direction are constrained by the streets of a city. At the beginning, MNs are at certain points on the streets and then select respective destinations. An MN moves toward the destination along a calculated path and at a selected speed (most often it is a shortest path) from the source point to the destination point. After arriving at the destination, it stops with a defined pause time and then repeats the process for the next selected destination. The city mobility model is much closer to real applications in an urban area. There are many variations and additional parameters included in the city mobility model to make it realistic, such as adding a random pause time at intersections, having finite acceleration, deceleration, and so on.

The ***group mobility model*** is different from other models that assume that MNs move independently. The group of nodes has a similar mobility pattern, such as speed and direction. However, for each MN, speed and direction may be slightly different. One of the first examples of the group mobility model is the *exponential correlated random mobility model* [9], given by:

$$s(t+1) = s(t)e^{-\frac{1}{\delta}} + \sigma\sqrt{1 - e^{-\frac{1}{\delta}}}r,$$

where $s(t)$ denotes the speed, δ represents the adjustment in the speed, and r indicates the random Gaussian variable with variance σ. Each mobility model incorporates some specific features. So, when investigating the heterogeneous connections of MNs in a MON, one or more mobility models may be used according to the application specifications.

10.2.2 Connection Process

Based on the mobility model(s), the essential problem in a MON is the connection duration and the frequency of an MN. Similar to the process model given in [10], the connection can be used to describe the attributes between a pair of MNs in a sequence of time slots.

As delivery latency is the biggest concern in a network, the model is based on the time duration between two consecutive connections. Given the state transition diagram in Figure 10.2, we define the status of a connection. During time slots $t = 0,1, \ldots$, connection process $L_t^{(u,v)} = 1$ if MNs u and v are connected. Otherwise, $L_t^{(u,v)} = 1$. For the sequence of time slots $t = T_0^{(u,v)}, T_1^{(u,v)}, \ldots, T_k^{(u,v)}, \ldots$, $(T_0^{(u,v)} < T_1^{(u,v)} < \ldots, T_k^{(u,v)} < \ldots)$ such that $L_t^{(u,v)} = 1$, time $\Delta T_k^{(u,v)} = T_{k+1}^{(u,v)} - T_k^{(u,v)}$ is the interconnection duration after the kth connection. Every connection duration is assumed to be independent and identically distributed (i.i.d.). Then, interconnect duration ΔT can be identified by:

$$P[\Delta T \geq t] = t^{-\alpha},$$

where α is the power law coefficient with $\alpha > 0$.

Remaining interconnection time is defined as:

$$R_t^{(u,v)} = \min\{t' - t \mid t' \geq t \wedge L_{t'}^{(u,v)} = 1\}.$$

End-to-end latency from source u to destination v denoted by $D^{(u,v)}$ is the accumulation of the remaining interconnection time for every hop, which can be computed by:

$$D^{(u,v)} = R^{(u,r_1)} + R^{(r_1,r_2)} + \cdots + R^{(r_i,v)},$$

where $r_1, \ldots, r_i$ is a sequence of MNs acting as intermediate nodes. While the network mobility model and the number of MNs in the network are determined, the sequence of intermediate nodes leads to more remaining interconnection time. With different selection strategies, the sequence of intermediate nodes will be different, and hence the expected end-to-end latency $E[D(u,v)$ also changes.

It is obvious that a higher value of parameter α means shorter interconnection duration. If α is above a certain threshold, the expected end-to-end message latency could go to infinity. To overcome this problem, the source MN may forward m copies of the same message to different intermediate nodes. Then, end-to-end latency can be reduced further since the latency is governed by the minimum end-to-end delay among all m copies.

10.2.3 Heterogeneous Connections

Heterogeneous MNs (i.e., MNs with heterogeneous interfaces) can benefit from the ability to access multiple networks. Due to the decreasing price of wireless interfaces, an MN is being equipped with multiple heterogeneous network interfaces for possible connections to technologies such as Bluetooth, WLAN, GSM, 3G, and so on. A number of accessible networks increase the connection opportunities for the source MNs, while intermediate MNs with multiple heterogeneous interfaces can act as network bridges between different networks. Messages can be selected to be delivered to different networks based on performance factors such as message delays, delivery probability, connection duration, and so on. It is also possible to maintain the upper-layer connection with continuous handover in different networks.

An increase in the opportunities for access with heterogeneous networks lies in the following aspects. First, the network node density is increased. With access to multiple networks, the whole network consists of various types of nodes. The number of nodes will be much larger than in a single network. In other words, network node density is increased to have the corresponding effect on both connection probability and delivery probability.

Second, an MN has a better chance of having access as it may be served by multiple networks. As discussed above, a wireless device on a vehicle may be covered by a vehicular ad hoc network (VANET) and a cellular network at the same time. If the device is equipped with only a single VANET interface, the chance of connection is only between vehicles. If the device is equipped with two interfaces, it can access either or both of the networks. In the example, the reverse direction vehicles have shorter connection duration, while they can utilize a cellular network to enhance the message delivery rate. If a vehicle does not have a cellular network interface, it can send messages to another vehicle that has the cellular interface and move in the same direction. If the vehicle itself has the interface and is able to reach a cellular tower, it can send the messages directly to the cellular network. In a similar way, the messages can be delivered from a cellular network to a VANET until it reaches the destination.

Third, the duration of a connection can be improved. In a single network, intermittent connection always leads to interruption if it has to use store-carry-forward technology to forward messages. If multiple networks are available, chances of finding a shorter delay or a robust route is increased. With the handover among different networks, a relatively stable connection can be built intermittently on MAC links.

As compared to the wireless resource cost of message duplication, the additional cost of multiple interfaces is relatively low. The obvious advantages of multiple interfaces enable MNs to hand over among different networks, prolong connection duration, and add opportunities to have better QoS.

10.2.4 Network Capacity

Because connectivity is not always present, message transmission is based on a store-carry-forward method. This means that messages are stored at an MN when the connection is not available and are forwarded when the connection becomes feasible. It is then possible for the capacity of the network to exceed the capacity bounds of traditional wireless networks. We use an example to illustrate the capacity gain as a result of the mobility and the special forwarding method.

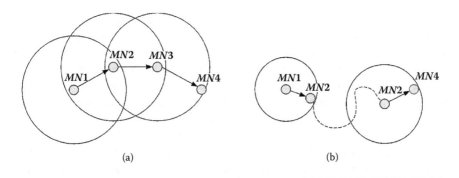

Figure 10.3 Capacity gain.

As shown in Figure 10.3, the message transferred from MN1 to MN4 has to go through three hops in a conventional transmission method. To avoid interference, each transmission has to occupy a region $S_I = \pi r_I^2$ where no other transmission is allowed. This means that the MNs in the region of S_I will suffer from interference from MN1. We denote this region as the interference space cost of the transmission from MN1 to MN2. Three-hop transmission leads to $3S_I$ space cost, with the assumption that each MN is using the same transmission power. In a MON, MN1 is able to transmit messages to MN2 at a lower power level since MN2 is closer to MN1. After MN2 receives the messages from MN1, it has the chance to move toward the destination MN4, as the dotted line shows in the figure. Once the connection between MN2 and MN4 is made, the message can be delivered from MN2 to MN4 directly. The space cost due to interference is less than $2S_I$ in this manner. So, the network capacity can benefit from the mobility of MNs.

Analytical work on capacity enhancement as a result of mobility has been done by [4], which shows that the average long-term throughput per source destination pair can be kept constant, even if the number of nodes in a unit area is increased. That is very different from the results in a fixed network, and a dramatic performance improvement can be obtained. Of course, the increase in capacity is at the cost of delivery latency.

10.3 Network Selection

In traditional wireless networks, MNs are connected to only one type of network. For example, consider a cellular phone, which is designed for connecting with only one type of network. Each mobile device is designed for connection with a cellular base station among multiple cells. It should authenticate itself and get permission from the base station before it can actually transmit packets to it. This process is followed by every cell to which it connects. That is, it should be authenticated and receive authorization (AA) for every cell along its motion path. The time required for each AA process produces a disruption in service while traveling in a fast-moving vehicle. The connection is completely lost if an MN moves out of the perimeter of a cell and there is no other cell to continue the service. Therefore, it is a common experience to face disruption or no service while traveling in a fast-moving vehicle. Mobile network planners generally place more cells in a densely populated area than one with sparse population. This kind of situation is encountered with all devices equipped with a single kind of terminal. The problem of disruption and dropped service can be overcome through a MON. The MNs in a MON may enhance connectivity by simultaneously connecting to multiple networks and may be able to perform seamless handover in case of nonavailability of one type of network along its path of mobility. However, for simultaneous heterogeneous network connection, the MNs in a MON should be multimode terminal devices capable of connecting to different types of access technologies. This can be achieved through enhancement in the software-defined radio technologies and is a research field for hardware engineers.

In this chapter, we concentrate on the techniques that can be used to improve connectivity in MONs, assuming the presence of multimode devices. A mobile device with a multimode terminal should be able to detect the presence of multiple networks in its vicinity and should be able to connect to them simultaneously, depending on the requirements of the users. So, while traveling in a fast-moving vehicle, the MN will be able to detect the strength of the received signals from different networks and decide on the handover point while redirecting ongoing service to another network before moving out of the vicinity of the one in use. In the case of total network outage, the device should be able to detect such noncontinuity prior to an actual loss and inform the user about potential disruption of service. Detecting a network outage and notifying the user is trivial. However, the process of network discovery, selection, and handover is complicated and needs detailed investigation. In this section we discuss the challenges encountered in network discovery and network selection and some solutions to address the issues.

10.3.1 Network Discovery

Network discovery is the preparatory phase for network selection, where the parameters gathered are used to calculate certain acceptance criteria. As there can be numerous parameters in the network, considering all of them makes the process complicated and time consuming. For simplicity, based on several factors, they can be narrowed down to six broader categories: availability, throughput, timeliness, reliability, security, and cost [11]. For example, received signal strength and coverage area have been placed under availability. Bandwidth and service rate have been grouped under throughput. Timeliness can be divided into delay, response time, and jitter. BER, interference, burst error, average number of retransmissions per packet, and packet loss ratio come under reliability. Cost can be separated into price and power consumption.

Network discovery involves two steps [12]. First, the network elements must be identified ahead of time, which consists of determination of the availability of access points through beacon signals and existence of routers, DHCP servers and other kinds of servers like AAA servers, Session Initiation Protocol (SIP) servers, and so on. After this determination is made, mobile devices need to set up connections in advance to communicate with these networks. Communication between mobile devices and networks include exchange of messages for information about the networks, authentication, authorization, and getting an IP address. Network discovery can be of one of the following types: centralized, where the networks provide information about the number of radio access networks (RAN) available; and distributed, where the terminals search for any available RAN or hybrid, which is a combination of both.

10.3.1.1 Network-Based Network Discovery

In network-based access discovery, the network operators manage the loads in their network and aim to optimize throughput. This has been successfully implemented in 2G and 3G systems. However, network-based discovery and selection in a MON requires the presence of a third-party agent. It has often been noted that a huge amount of information generated from the network and the user preferences could overwhelm the capabilities of the third-party agent. Also the QoS information of the wireless part is of more relevance than the entire network.

10.3.1.2 Terminal-Based Network Discovery

Terminal-based network discovery concentrates on the requirements of the user applications. Networks allocate different resources to different users depending on the context of their application programs. When a network discovery is needed, the mobile device turns on all the terminals for available networks and sends a resource query (RQ) to all of the neighboring access points (APs) to find out the resources that they can allocate to the application. A typical RQ looks like the following:

$$RQ = user(application,\ param1,\ param2,...,paramN),$$

where user parameters are the required information for radio resources in the network and N is the number of parameters. For example, param1 may be available bandwidth; param2 may be delay, and so on. Once the MN receives all the information it requires, it then decides on the selection of the network that best suits its needs. Terminal-based network discovery brings minimal changes to the existing network infrastructure. Only two messages are added between the terminal and the networks. It is not only flexible and easy to implement, but it also fully considers all user preferences. Hence, this is an economical way of network discovery.

10.3.1.3 Hybrid Approach to Network Discovery

The hybrid approach focuses on network discovery in both homogeneous and heterogeneous networks and handover for mobile and stationary users. Handover for mobile users is used when the link condition changes in the network. Stationary users may decide to hand over when there is any change in the service requirements or there is the presence of another network with a better QoS offering.

10.3.2 Network Selection Algorithms

Network selection algorithms can be categorized into two types: fuzzy logic–based schemes and multi-attribute decision making schemes.

10.3.2.1 Fuzzy Logic–Based Scheme

Fuzzy logic–based network selection schemes are mainly designed to maintain QoS in the network terminals. They consist of a group of fuzzy logic rules in the form of If–Then statements and are used for decision making. The system has a fuzzy logic controller that accepts users based on network-defined attributes and produces a ranking matrix of all available networks. The network with the highest rank is selected by the user. The procedure is quite simple, as each network advertises its characteristics through periodic beacon signals. User terminals can maintain a table of these characteristics. Composite ranking of the networks can be done by assigning equal priority to the networks or by prioritizing the attributes to satisfy the requirements of the user. Such a formally produced network is called *network-specific ranking* and is done in [11] following the Mamdani method [13]. In user-specific ranking, the users can specify different criteria of their choice for selecting a network (Figure 10.4).

Human experience and knowledge is represented by fuzzy logic as fuzzy rules. Hence, under some weak assumptions, they can provide fairly accurate results.

The rules in a fuzzy logic scheme need to be manually configured by the user before inputting into the controller. This could become cumbersome and complex as the number of attributes increases. However, a MON is highly diverse in access technologies. Hence, the scalability of fuzzy logic schemes is low when applied to a MON.

10.3.2.2 Multi-Attribute Decision Making Scheme

This scheme consists of utility functions that take inputs about the link quality of each radio network and the user preferences and create a ranking among the networks based on the expected utility. The decision is performed based on two scenarios: a downgrade in the connection and the corresponding upgrade. A downgrade occurs when an existing connection is lost and the user has to choose a network with lower QoS than the existing one. When a network with more promising

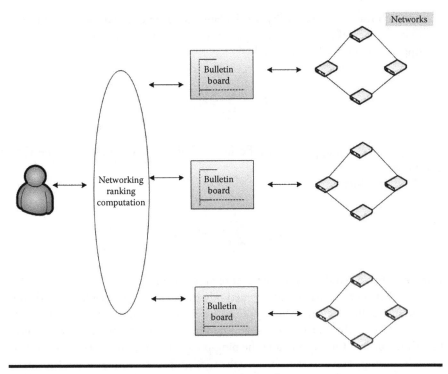

Figure 10.4   Users select a network from the rank computation.

features comes up, the user may decide to terminate an existing connection in favor of enhanced services, which is called an upgrade of connection. So, the system not only needs to monitor link conditions periodically, but also ought to predict if a network will be available $P_{avail}(N_a)$ in the near future. To perform a downgrade, the user has to first select the network with the highest expected utility according to the following relation [14]:

$$U_{expected}(N_i) = U_{total}(N_i) \times P(N_a).$$

Total utility is derived from the link conditions and user preference. Upgrade is performed if a previously inactive network becomes available and it has the highest expected utility. Total utility can be calculated from a number of mappings [15] as follows:

Quality-to-resource mapping: This is a mapping of the QoS level of a traffic flow to the resources required for maintaining that QoS level.

Quality-to-utility mapping: This is a mapping of the QoS level of a traffic flow to the user utility related to that quality.

Resource-to-cost mapping: This is a mapping of the cost to the resources. This helps to fix the pricing scheme for each access network.

For each user, the total utility is calculated as a sum of the utility derived from the QoS of the traffic flow for each application minus the cost to achieve the QoS. Application utility is expressed by:

$$U_i = \sum_{j=1}^{f_i} u(q_{i,j}),$$

where $q_{i,j}$ is the mapping of traffic flow f_i of application A_i to some QoS parameter j.

Net user utility is given by:

$$U_{total} = \sum_{i=1}^{N} U_i - c(q_{i,j}),$$

where c is the cost and N is the number of applications.

Multi-attribute decision-making methods are flexible and scalable. Users do not need to manually set them, which reduces the occurrences of errors in the decision-making process. Multi-attribute decision-making methods do not take into account user distribution. They also ignore the ping-pong effect that may occur due to network switching. Hence, even for a slight QoS improvement, they tend to make users pay more for the service.

10.4 Message Forwarding

Unlike message forwarding in a conventional network, the packet forwarding in an opportunistic network is based on a store-carry-forward mechanism. Because the connection to neighboring MNs is not always available, the message has to be stored at the MN before the connection becomes available. The MN may move a certain distance before a node is able to forward messages to a next-hop MN. Such procedures are common in opportunistic network transfer and lead to its special features as follows.

As the connection to a next-hop MN may not be available, an intermediate MN has to store the message. While the duration of the connection between a pair of MNs is long, one of the MNs may receive a large volume of messages from others, since the receiver may not be able to forward the message due to the absence of a next-hop MN. Thus, the requirement of buffer size under such situations will be much larger than that in a conventional network. Let us assume the message arrival at the receiver has a Poisson distribution with parameter λ and the forwarding rate is μ, with an exponential distribution for message forwarding process time. In a conventional network, the queue length is approximated by $\frac{\lambda}{\mu-\lambda}$ (according to the queue length computation of M/M/1 queue). In a MON, if forwarding is not available, the required buffer size at the receiver is λT, which is linearly increased with transmission duration. In most cases, $T \gg \frac{1}{\mu-\lambda}$, where $\gg$ means *much larger than*. So, the hardware storage problem has been considered and the requirement is much higher for MONs.

While the messages are stored, the MN may move from one place to another and may establish a connection at a new place. During this period, the carried messages also "move" the same distance. This is very different from a conventional network. Even in the area of mobile ad hoc networks (MANETs), the messages are assumed to be stored at the MN for a negligible duration. So, the movement of an MN during transmission has a very small effect and is not counted in a MANET. In contrast, the movement distance may be long during the storage in a MON. If the movement is toward the destination and it moves the messages closer to the destination, the moved distance should be counted as a delivery distance. Because the velocity of an MN is much lower than the wireless signal propagation speed in the air, the delay for the message to be carried by the MN from one place to another place is much longer than it takes for the message to be transmitted the same distance.

Besides single-copy forwarding, due to long delays and low probability of successful delivery, MNs in MONs may forward multiple copies of a single message. It is similar to multi-path routing. The difference is that the same messages are delivered on multiple intermediate MNs in MONs. While one of the copies reaches the destination, the message is thought to be successfully delivered. Conventional TCP is not suitable for MONs, as a broken connection may last for a longer time and could happen frequently. The overhead of TCP may be excessive and therefore meaningless as a measure to be used.

To deal with these special aspects in MON forwarding, many message forwarding approaches have been introduced and can be divided into two categories: single-copy and multiple-copy forwarding. Only one message copy is forwarded in the network in single-copy forwarding. The message is either in the air during the transmission or is carried by an MN. In contrast, in a multiple-copy forwarding strategy, more than one copy of a single message exists in the network at the same time. When one of the copies reaches the destination, the message delivery is said to be successful. Obviously, multicopy forwarding has more chance to deliver a message to the destination within a shorter time. However, it requires much more in wireless resources, storage space, and energy consumption. The selection and usage of strategies are tightly related to applications.

We show these state-of-the-art forwarding approaches from a simple to a complex one. To have an intuitive sense of these approaches, we illustrate the performance and impact of different approaches by analyzing them using the same example.

10.4.1 Single-Copy Forwarding

Single-copy forwarding has naturally evolved from conventional MANETs. The difference is that it has to deal with disconnected situations. Conventional

MANET forwards messages to the next-hop MN only with an existing routing path to the destination. Any next-hop MN without such a path is not considered by the source MN. Since such a next-hop MN having a routing path to the destination is assumed not to be available in a MON, the message has to be forwarded to a next-hop MN that does not have a routing path to the destination at that time, but that is able to carry and move the message toward the destination in the near future. So, successful message delivery depends on if the MN can forward messages to a correct next-hop MN. As conventional MANETs evaluate the routing path with some metrics for shorter delays, higher throughput, and so on, single-copy forwarding in MONs mainly depends on a per-hop forwarding decision based on selecting a next-hop MN with higher delivery probability to the destination.

As a method for saving network resources, single-copy forwarding approaches have drawn much attention in the development of MONs. The evolution of single-copy forwarding approaches is illustrated as follows.

10.4.1.1 Direct Forwarding

This is a very simple transmission approach. If MN u has a message destined to MN v, it sends the message if and only if the distance between u and v are within the transmission distance r_T. Otherwise, MN u stores the message. Implementation of this approach is simple. While u and v are within the transmission distance, MN u checks the address of the connected MN. If the address matches any address of the stored messages, it forwards the corresponding messages to the connected MN.

An example of this implementation is an application for monitoring wild animals, where the abstraction of the network is shown by Figure 10.5. The sensors installed on the wild animals collect sensed information and store it at the sensors. Base station w, which collects data from sensors, can be placed at the locations that the wild animals are most likely to visit. At the beginning, shown in Figure 10.5(a), MN u is not within the transmission region, but it has a message to transmit. So, it stores the message. When it moves into the transmission region of the destination w, denoted by the dashed circle, the message forwarding can be performed, as shown in Figure 10.5(b).

However, in such an approach, the delay largely depends on the mobility pattern of MNs. The problem with this method is that messages may not have any chance of being delivered in some cases. For example, if the wild animals migrate to some other area and never return to the place where the base stations are located, the data can never be collected. So, for many applications, this simple approach is not appropriate. Otherwise, in general, the delay is unpredictable or unbounded.

10.4.1.2 Randomized Forwarding

The major problem of direct forwarding is that it can only have the transmission within the region of the destination, which could be very limited. It should take

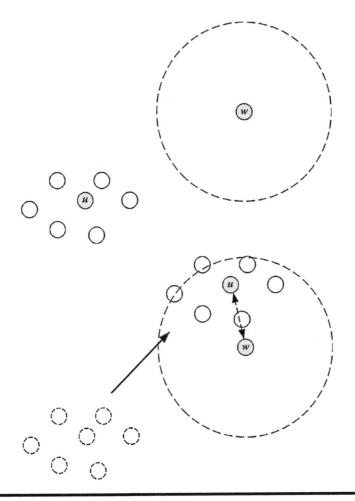

Figure 10.5   Direct forwarding.

advantage of wireless multihop transmission as well. So, the source MNs do not have to pass the transmission region around the destination. Randomized forwarding is a simple approach to achieve this and does not need any knowledge of network topology. For a source MN u, it forwards the message to the encountered MN with probability p, where $p \in [0,1]$.

Consider the example of wild animals again. As shown in Figure 10.6, the group of animals may not go through the transmission region of w. However, when a group of animals meets another group of animals with wireless sensors, the messages could be delivered with probability p. For example, the transmission can be performed from u to v, as shown in Figure 10.6(a). When v moves within the transmission region of w, the messages can be sent to the destination.

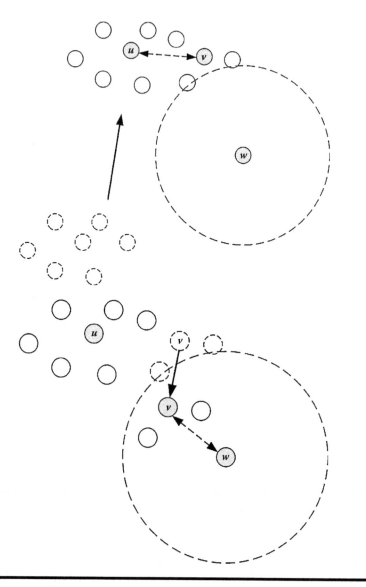

Figure 10.6 Randomized forwarding.

Although this approach is still naïve, it shows several obvious advantages as compared to direct forwarding. First, the effective transmission can be performed in a
much wider region due to multihop message relay. Second, source MNs do not have
to move toward the destination. The delivery can be performed with the help of other
MNs moving toward the destination. Thus, it has a higher delivery probability and
a shorter delivery delay. However, new issues do appear. One of the disadvantages
is that it may cause local loops. The message can be sent from *u* to *v*, but it finally

returns back to *u* after relaying through several hops. The transmission does not have a mechanism to record the MNs that it has passed. In addition, the message may be forwarded to a next-hop MN that has a lower probability of reaching the destination. For example, MN *u* is moving toward the destination. Before it moves close enough for transmission, MN *v* connects with MN *u* but moves away from the destination. MN *u* will forward the message to MN *v* with probability *p*. Then, if the message is delivered form *u* to *v*, the performance is worse than with direct forwarding.

10.4.1.3 Mobility-Based Forwarding

Due to the blindness of randomized forwarding, other approaches seek to utilize certain attributes of the MNs in the network and ensure message forwarding toward the destination with a higher probability. Mobility-based forwarding [16] uses mobility information to select an intermediate MN. The selection and forwarding of messages is based on the knowledge of MN mobility. For MN *u*, the effective mobility of *u* toward *v* is computed by the following method. Let $\vec{V}_u$ denote the velocity vectors of *u*. An auxiliary vector from source MN *u* to destination MN *v* is denoted by $\overrightarrow{uv}$. Then, an angle θ is computed by:

$$\theta = \cos^{-1}\left(\frac{\overrightarrow{uv}\vec{V}_u}{|\overrightarrow{uv}||\vec{V}_u|}\right).$$

Then, the effective velocity of MN *u* is the projected velocity on $\overrightarrow{uv}$, which is:

$$V_{ef}(u) = \sin\theta \overrightarrow{uv}.$$

If θ is smaller than $\frac{\pi}{2}$, MN *u* is moving closer to MN *u*. Otherwise, MN *u* is moving away from MN *v*. With the computation of the effective mobility of connected MNs, the source MN predicts the movement of MNs within transmission distance and selects one of them and forwards the messages. This method sends messages to an MN that will most likely move toward the destination. It can perform better than the direct transmission method in most cases.

Let us still take the example of the direct transmission approach to show the improvement in the message delivery. While wild animals move away or migrate from locations around the base stations, the messages still can be forwarded to the wild animals carrying sensors and moving toward the base stations once they are within the transmission distance.

However, several assumptions are relatively weak when implementing this approach in some applications. The method is based on the assumption that the direction of movement would not change. If the intermediate MN often changes its speed and direction, the message delivery still suffers from a long delay. In the worst case, the message may never arrive at the destination. The approach also assumes that the MN knows its location, the destination's location, and the locations of intermediate MNs. For MNs equipped with GPS, this is not a big deal, but this is not always true for all MNs.

10.4.1.4 Location-Based Forwarding

The main concept of the location-based approach is to have messages delivered to MNs closer to the destination. The fundamental problem is to define the location relationship for the MNs. Since they keep moving in the network, the definition of the location relationship between MNs must also consider the changes in location of MNs. Euclidean virtual space is one efficient method that can be used for this purpose.

Several reference nodes are defined in the network. The location of an MN can be identified by a relationship to these nodes. When MN passes by one of the reference nodes, the connection between them can be recorded. So, the distance is defined to reflect the probability that an MN passes a reference node. A longer distance between an MN and a reference node means the connection probability is lower. Given a set of reference nodes, denoted by $\mathcal{R}$, the probabilities of an MN visiting around the place of reference nodes can be identified. So, the scope of the MN's location can be characterized.

After the locations of MNs in Euclidean space are determined, the evaluation of the distance between two MNs can be done using one of following methods [17].

Euclidean distance: This is the root of the sum of the square differences between the two coordinates. For two MNs, and v, and total N reference nodes, the Euclidean distance between u and v is expressed by:

$$d_{uv} = \sqrt{\sum_{k=1}^{N} (x_{uk} - x_{vk})^2},$$

where x_{uk} is the kth weight of the kth reference node.

Canberra distance: This examines the sum of a series of fractional differences between the coordinates of a pair of MNs. The expression is:

$$d_{uv} = \sum_{k=1}^{N} \frac{|x_{uk} - x_{vk}|}{|x_{uk}| + |x_{vk}|}.$$

Cosine angle separation: This is used to represent the cosine of the angle between two MN locations. Higher cosine angle separation indicates that the locations of two MNs are similar, where the computation is as follows:

$$s_{uv} = \frac{\sum_{k=1}^{N} x_{uk} x_{vk}}{\sqrt{\sum_{k=1}^{N} x_{uk}^2 \sum_{r=1}^{N} x_{kr}^2}}.$$

Matching distance: This counts the number of location weights that are within a given threshold δ. The equation for computation is as follows:

$$n_{uv} = \sum_{k=1}^{N} \{|x_{uk} - x_{vk}| \le \delta\}.$$

These four methods focus on different aspects of the locations of two MNs and their changing attributes. It is hard to qualify which one is better without knowledge of the mobility model of MNs. One or more of the methods above can be used to estimate the similarity between two MNs in the network.

The strategy of message forwarding in this approach is to forward messages to an MN that is at a closer distance. The convergence speed of the messages to the destination (i.e., the delay from the source to the destination) highly depends on the mobility model of MNs, which is worth further investigation. The usefulness of the approach is based on the assumption that intermediate MNs acknowledge the location of the destination. So, an efficient information exchange is desirable.

With this approach, the messages can be forwarded toward the destination with higher probability. The efficiency will be increased if the location estimate is accurate. However, in some cases the representation of MNs in Euclidean space may not accurately characterize the mobility of the MNs. For example, two MNs visit the same location in the same way, but with certain delays between them. They will have the same representation in Euclidean space but may never be connected. A study should be conducted before implementing this in a network.

As considered in the example of wild animals, base stations located at different places for data collection can be used as the reference nodes. The sensors on the animals can estimate the locations in Euclidean space formed by the base stations. The messages are delivered toward one of the base stations.

10.4.1.5 Probability-Based Forwarding

Probability-based forwarding [18,19] forwards the messages about the probability that the next MN can deliver the message to the destination. Each MN maintains the delivery predictability to other MNs. For example, the delivery predictability from MN u to MS v is denoted by p_{uv}. This assumes that two MNs will meet again with higher probability if they meet very often. Otherwise, the probability is low. So, the value of delivery predictability is updated regarding the two situations. When two MNs encounter each other, a new value is updated by:

$$p'_{uv} = p_{uv} + (1 - p_{uv})P_c,$$

where P_c is a constant for the update weight.

On the other hand, if MNs u and v do not meet for a duration, the delivery predictability should be reduced. So, the decrease of the value can be computed by:

$$p'_{uv} = p_{uv}\gamma^{\tau_u(v)},$$

where γ ($\gamma \in (0,1)$) is a positive aging constant and $\tau_u(v)$ denotes the elapsed time since the last connection time of MNs u and v.

One of the features in this approach is the transitive property of delivery predictability. Knowing delivery predictability p_{uv} at MN u and delivery predictability p_{uv} at MN v, the delivery predictability p_{uw} at MN v can be updated by:

$$p'_{uw} = p_{uw} + (1 - p_{uw})p_{uv}p_{vw}\theta,$$

where θ is a parameter to decide the impact by transitive estimate.

Based on delivery predictability, an MN sends the message to a connected MN if and only if the delivery predictability of the connection MN to the destination is higher. The benefit of the probability-based approach is apparent. The delivery probability can be increased and the end-to-end delay can be reduced because it sends messages to MNs with a high probability of reaching the destination. As compared to the native epidemic forwarding approach, the overhead is that the MNs in the network have to exchange delivery predictability information and maintain updated information.

10.4.1.6 Cost-Based Approach

The selection of forwarding MNs depends on the estimate of delivery likelihood in [20]. MN u keeps track of the probability of meeting MN v. The frequency f_v^u is used to denote the likelihood that u meets v next time. For every time that u meets v, the value of f_v^u is updated to reflect the frequency of meeting. It is based on the assumption that if the MNs met with higher frequency in the past, they will also meet with high probability in the future. When MNs meet each other, they also exchange delivery likelihood information.

After a period of exchange, an MN can have information about the delivery likelihood to other MNs. It can be used for the computation of routing cost, which is

$$c(u,.,w,.,v) = \sum_{k \in P_{uv}} (1 - f_k^u),$$

where P_{uv} denotes the path from u to v. MN u will select the next MN on the minimum cost path.

10.4.1.7 Utility-Based Approach

As discussed in [17], more knowledge gathered by MNs can generate a better chance for successful delivery of messages to the destination. The knowledge includes the mobility pattern, storage capacity, and so on. By utilizing sufficient information, appropriate prediction can be made by exploring a knowledge-based approach. Context-aware routing (CAR) is one such approach, introduced by [21]. Context is defined as the set of attributes related to message delivery. The computation of delivery probability is based on the prediction of the attributes for the context.

Context information is related to MN v in the network. The attributions are the elements in a set, denoted by $\mathbf{A}^v = \{A_1^v, A_2^v, .., A_N^v\}$. Because the attributes may not be updated on time and may not completely reflect the current status of MN in the network, each MN performs Kalman filter prediction to estimate the current status of other MNs. One of the features is that the Kalman filter is a recursive filter without a requirement for storing history information, which decreases the storage space of the MN. With the predicted attributes, denoted by $\mathbf{a}^v = \{a_1^v, a_2^v, .., a_N^v\}$, a utility function is used to represent the delivery probability to other MNs in the network. The utility function is computed by the summation of the weighted attributes. MN u forwards the messages to an MN that has a higher utility to the destination MN.

The example of wild animals can also be used to implement a cost-based or a utility-based approach. Data messages from a sensor on an animal can be forwarded to the sensor on another wild animal that has a lower cost or higher utility of delivering data messages.

Mobility-based, location-based, probability-based, cost-based, and utility-based forwarding approaches all try to have the message delivered at the destination, but performance highly depends on the mobility model of the applications.

10.4.2 Multicopy Forwarding

Increasing the number of copies forwarded to different MNs can effectively increase the delivery probability and reduce delivery delay. The problem is how many copies should be sent to other MNs and what kind of performance can be expected? Multicopy forwarding can be generally divided into following categories: epidemic forwarding and utility-based forwarding.

10.4.2.1 Epidemic Forwarding

The spray and wait scheme in [5] is similar to direct forwarding with a single copy. The source MN sends m copies to the m distinct MNs during its movement. These m MNs carry and transmit the messages to the destination once they meet the destination MN. A multihop epidemic forwarding is illustrated in [22]. Duplicated messages carried by an intermediate MN can be sent to a next intermediate MN, rather than only the destination. Since it is not a direct transmission, several additional procedures are needed to enable the transmission using multiple hops so as to improve the performance. When MN u has the connection with MN v, it first transmits a summary vector to v, which is a compact representation of all the messages buffered at u. Then, v performs a logical operation to check the difference in stored messages at u and v. After that, v sends a request message for the transmission of unknown messages.

For multicopy forwarding, the storage problem becomes important. Multiple copies of the messages require a larger buffer size for a roaming MN. When the buffer size is limited, MN has to make policy and decide which message is to be dropped when the buffer overflows. For example, hop count is used in [22] to record the number of hops traveled by a message, which is similar to the TTL field in an IP message. If the value of hop count is reduced to one, the MN will hold the message until it meets the destination of the message, unless the message is erased due to a shortage of memory. When the required message storage is more than the size of the buffer, some of the messages have to be dropped. The hop count is also used to decide which message will be dropped. Messages with a lower hop count will be dropped first, which requires that the priority of messages can be controlled by setting the initial hop count value. It is worth noticing that when the hop count of all messages is set to two, the approach is reduced to the spray and wait approach.

With epidemic forwarding, the source sensors on the wild animals will send the message to the first m other sensors it meets. Then, with the spray and wait method, the m carrier will store and send the message to the destination while it is within the transmission region of the destination. Alternatively, in epidemic forwarding [22], each of messages will be forwarded nearly independently with a specified number of copies.

10.4.2.2 Utility-Based Forwarding

The utility-based approach [5] can improve the performance during message forwarding of intermediate MNs. In the multihop epidemic method, messages will be forwarded to the connected MN that does not have the message copy. However, it may not be efficient. For example, the message carrier MN may be moving toward the destination and the connection MN is not. Under such a situation, the delivery of a message between two MNs will lead to an unexpected result. Thus, the utility-based approach uses the estimate of the utilities of MNs to solve this problem. Utility can be built based on the elapsed connection time of two MNs. For MNs u and v, the elapsed time since the end of their connection can be denoted by $\tau_u(v)$. The utility function of MN u can be built taking into account the MNs that have met. The mapping of utility by MN u for MN v is represented by $U_u(v)$. So, MN u will forward the message to MN w with a destination of MN v, if and only if:

$$U_w(v) > U_u(v) + \beta,$$

where β is the threshold used to avoid the ping-pong effect in forwarding.

During multicopy forwarding, the utility-based approach consists of the following two phases. In the first phase, the source MN sends m copies of the original message to the first m distinct MNs to which it can connect. In the second phase, for each MN carrying the message, it forwards the message to a next MN that has a utility higher than the threshold.

Utility-based forwarding enables the sensors in the wild animal example to select m next-hop MNs for the message copies. It can very effectively improve delivery rates and lower delays. However, it is at the cost of network wireless resources in terms of bandwidth, buffer, energy consumption, and so on. If the battery and memory on sensors are limited, the cost of these aspects has to be carefully considered in the application.

10.5 Conclusion

Connection enhancement in a MON is to optimally select a set of networks and a set of MNs for per-hop delivery. The objective of connection enhancement is to increase message delivery and reduce delivery latency with various constraints in terms of wireless resources, storage space, energy consumption, and so on. As we discussed in this chapter, mobility models and heterogeneous connections have a fundamental impact on the performance of a MON, and various strategies are detailed for network connection selection and message forwarding. Heterogeneous connections increase connection opportunities and require appropriate network connection selection. Message forwarding is used to appropriately select a set of intermediate nodes with certain strategies. For each of them, we presented the state-of-the-art approaches with basic procedures and associated limitations.

References

[1] K. Fall. A delay-tolerant network architecture for challenged Internets. In *Proceedings of the Conference on Applications, Technologies, Architectures, and Protocols for Computer Communications*, pp 27–34, 2003.

[2] P. Juang, H. Oki, Y. Wang, M. Martonosi, L. S. Peh, and D. I. Rubenstein. Energy-efficient computing for wildlife tracking: Design tradeoffs and early experiences with ZebraNet. *ACM SIGPLAN SIGOPS operating systems review*, pp 96–107, 2002.

[3] T. Small and Z. J. Haas. The shared wireless Infostation model: A new ad hoc networking paradigm (or where there is a whale, there is a way). In *Proceedings of the 4th ACM International Symposium on Mobile Ad Hoc Networking and Computing* (MobiHoc 2003), pp 233–244, 2003.

[4] M. Grossglauser and D. N. C. Tse. Mobility increases the capacity of ad hoc wireless networks. *Transactions on Networking* (TON), pp 477–486, vol. 10, no. 4, 2002.

[5] T. Spyropoulos, K. Psounis, and C. S. Raghavendra. Efficient routing in intermittently connected mobile networks: The multiple-copy case. *IEEE/ACM Transactions on Networking* (TON), pp 77–90, vol. 16, no. 1, 2008.

[6] T. Camp, T. J. Boleng, and V. Davies. A survey of mobility models for ad hoc network research. *Wireless Communications and Mobile Computing*, John Wiley & Sons, pp. 483–502, vol. 2, no. 5, 2002.

[7] J. Broch, D. A. Maltz, D. B. Johnson, Y-C Hu, and J. Jetcheva. A performance comparison of multi-hop wireless ad hoc network routing protocols. In *ACM/IEEE International Conference on Mobile Computing and Networking*, pp. 85–97, 1998.

[8] J. G. Markoulidakis, G. L. Lyberopoulos, D. F. Tsirkas, and E. D. Sykas. Mobility modeling in third-generation mobile telecommunications systems. *IEEE Personal Communications*, pp. 41–56, vol. 4, no. 4, 1997.

[9] M. Bergamo, R. R. Hain, K. Kasera, D. Li, R. Ramanathan, and M. Steenstrup. System design specification for mobile multimedia wireless network (MMWN) (draft). DARPA project DAAB07-95-C- D, 1996.

[10] A. Chaintreau, P. Hui, J. Crowcroft, C. Diot, R. Gass, and J. Scott. Impact of human mobility on opportunistic forwarding algorithms. Transactions on Mobile Computing (TMC), pp. 606–620, vol. 6, no. 6, 2007.

[11] S. Kher, A. K. Somani, and R. Gupta. Network selection using fuzzy logic. *International Conference on Broadband Networks*, pp. 876–885, 2005.

[12] Y. Chen and Y. Yang. Access discovery in Always Best Connected networks. In *Proceedings of the IEEE International Symposium on Knowledge Acquisition and Modeling Work*, pp. 794–797, 2008.

[13] J. Jang, C. Sun, and E. Mizutani, Neuro-Fuzzy and Soft Computing-A Computational Approach to Learning and Machine Intelligence, Prentice Hall, 1997.

[14] Q. Duan, L. Wang, C. D. Knutson, and D. Zappa. Autonomous and intelligent radio switching for heterogeneous wireless network. In *Proceedings of the IEEE Mobile Ad Hoc and Sensor Systems*, pp. 666–671, 2008.

[15] V. Gazis, N. Alonistioti, and L. Merakos. Toward a generic "always best connected" capability in integrated WLAN/UMTS cellular mobile networks (and beyond). *IEEE Wireless Communications*, pp. 20–29, vol. 12, no. 3, 2005.

[16] J. LeBrun, C. Chuah, D. Ghosal, and M. Zhang. Knowledge-based opportunistic forwarding in vehicular wireless ad hoc networks. In *Proceedings of the Vehicular Technology Conference*, pp. 2289-2293, vol.4, 2005.

[17] J. Leguay, T. Friedman, and V. Conan. DTN routing in a mobility pattern space. In *Proceedings of the SIGCOMM Workshop on Delay-Tolerant Networking*, pp. 276–283, 2005.

[18] A. Lindgren, A. Doria, and O. Schelen. Probabilistic routing in intermittently connected networks. *Lecture Notes in Computer Science*, 2004.

[19] A. Lindgren and A. Doria. Probabilistic routing protocol for intermittently connected networks. Internet draft, draft-irtf-dtnrg-prophet-00, February 2008.

[20] J. Burgess, B. Gallagher, D. Jensen, and B. N. Levine. Maxprop: Routing for vehicle-based disruption-tolerant networks. In *Proceedings of the IEEE International Conference on Computer Communications*, pp.1–11, 2006.

[21] M. Musolesi, S. Hailes, and C. Mascolo. Adaptive routing for intermittently connected mobile ad hoc networks. In *Proceedings of the IEEE International Symposium on world of wireless mobile and multimedia networks*, pp. 183–189, 2005.

[22] A. Vahdat and D. Becker. Epidemic routing for partially connected ad hoc networks. Duke Tech Report CS-2000-06, 2000.

Index